THE
GEOMORPHOLOGY
OF CANADA

AN INTRODUCTION

—

ALAN S. TRENHAILE

THE
GEOMORPHOLOGY
OF CANADA

AN INTRODUCTION

ALAN S. TRENHAILE

Toronto
OXFORD UNIVERSITY PRESS

Oxford University Press, 70 Wynford Drive, Don Mills, Ontario M3C 1J9

Toronto Oxford New York
Delhi Bombay Calcutta Madras Karachi Kuala Lumpur
Singapore Hong Kong Tokyo Nairobi Dar es Salaam
Cape Town Melbourne Auckland

and associated companies in
Berlin Ibadan

Canadian Cataloguing in Publication Data

Trenhaile, Alan S.
  The geomorphology of Canada

Includes bibliographical references.

ISBN 0-19-540791-1

1. Geomorphology — Canada. 2. Landforms — Canada.
3. Glacial landforms — Canada. I. Title.

GB428.5. C3T74 1990   551.4′1′0971   C90-094905-8

OXFORD is a trademark of Oxford University Press

  3 4 — 94

Printed in Canada

Contents

Acknowledgements

Figures 1.12, 1.13, 8.2, 8.8, 8.9. From *Hydrological Atlas of Canada* 1978. Energy Mines and Resources, Cartographic Information and Distribution Centre. Reproduced with permission of the Minister of Supply and Services Canada, 1990.

Figure 2.4. From L.C. Peltier, 'The geographic cycle in periglacial regions' in *Annals of the Association of American Geographers* 40 (1950). Reproduced by permission of the Association of American Geographers.

Figure 2.5. From J.K. Fraser, 'Freeze-thaw Frequencies and Mechanical Weathering in Canada' in *Arctic* 12:52 (1959). Reprinted with permission from the Arctic Institute of North America.

Figure 3.3. From M.J. Bovis, 'Earthflows in the interior plateau, southwest British Columbia' in *Canadian Geotechnical Journal* 22 (1985). Reproduced by permission of the Geological Society of America and M.J. Bovis.

Figure 3.5. In part from J. Béland, *Geological Association of Canada Proceedings* 8. Used by permission of the Geological Association of Canada.

Figure 3.7. From A.G. Lewkowicz, 'Periglacial systems' in D. Briggs, P. Smithson, and T. Ball, *Physical Geography* (1989). Reproduced by permission of Copp, Clark, Pitman Ltd, Toronto.

Figures 4.3 and 4.4. From G.S. Boulton, 'Processes and patterns of glacial erosion' in D.R. Coates, *Glacial Geomorphology* (1974), State University of New York, Binghamton, N.Y. Reprinted by permission of Donald R. Coates.

Figure 4.6 (c), (d), (e), (f). From *Quaternary Environments*, edited by J.T. Andrews, reproduced by kind permission of Unwin Hyman Ltd.

Figures 5.3 and 5.7. From *Géographie Physique et Quaternaire* 41 (1987). Reprinted by permission of Gaëtan Morin Éditeur Ltée.

Figure 5.4. From J.T. Andrews, 'The late Wisconsin glaciation and deglaciation of the Laurentide ice sheet' in W.F. Ruddiman and H.E. Wright, *North America and Adjacent Oceans During the Last Deglaciation* (1987). Geological Society of America, The Geology of North America, K-3. Reprinted by permission of John T. Andrews.

Figure 5.10. From P.F. Karrow and P.E. Calkin, 'Quaternary evolution of the Great Lakes' in *Geological Association of Canada*, Special Paper 30. Used by permission of the Geological Association of Canada.

Figure 6.1. From J.S. Scott, 'Geology of Canadian tills' in R.F. Leggett, *Glacial Till* (1976). Royal Society of Canada, Special Publication 12. Reproduced by permission.

Figure 6.3. From C. Hillaire-Marcel, D.R. Grant, and J.-S. Vincent, 'Comment and reply on "Keewatin ice sheet–re-evaluation of the traditional concept of the Laurentide ice sheet" and "Glacial rosion and ice sheet divides, northeastern Laurentide ice sheet, on the basis of the distribution of limestone erratics" ' in *Geology* 8 (1980). Reproduced by permission of the Geological Society of America and C. Hillaire-Marcel.

Figure 6.12. From *Journal of Glaciology* 20, 1978, pp. 367-92. Reproduced by courtesy of the International Glaciological Society and D.E. Sugden.

Figure 7.1. From S.A. Harris, 'Permafrost Distribution, Zonation and Stability along the Eastern Ranges of the Cordillera of North America' in *Arctic* 39: 34, 35 (1986). Reprinted with permission from the Arctic Institute of North America.

Figure 7.6. From S.A. Harris, 'Distribution of zonal permafrost landforms with freezing and thawing indices' in *Erdkunde* 35 (1981), pp. 81-90. Reprinted by permission.

Figures 7.7, 7.8, and 7.9. From H.M. French, *The Periglacial Environment* (1976). Reprinted by permission of Longman Group UK Limited.

Figure 7.11. From P.P. David, 'Sand dune occurrences of Canada, 1977'. Indian and Northern Affairs, National Parks Branch, Contract no. 74-230 Report. Reproduced with permission of the Minister of Supply and Services Canada, 1990.

Figure 8.6. From D.F. Ritter, *Process Geomorphology* (1978). Reprinted by permission of American Association of Petroleum Geologists.

Figures 8.11 and 8.12. From M. Morisawa, *Streams* (1968). Reproduced by permission of McGraw-Hill Publishing Company, New York.

Figure 8.13. From C.R. Neill, 'Measurement of bridge scour and bed changes in a flooding sand-bed river' in *Proceedings of the Institute of Civil Engineers (U.K.)* 30 (1965). Used by permission of Thomas Telford Publications.

Figure 8.14. From E.J. Hickin and G.C. Nanson, 'The character of channel migration on the Beatton River, northeast British Columbia, Canada' in *Geological Society of America Bulletin* 86 (1975). Reproduced by permission of the Geological Society of America and E.J. Hickin.

Figure 8.25. From K.J. Tinkler, 'Canadian landform examples – 2 Niagara Falls' in *Canadian Geographer* 30 (1986). Reprinted by permission of The Canadian Association of Geographers.

Figure 9.9. From L.D. Wright and A.D. Short, 'Morphodynamic variability of surf zones and beaches: a synthesis' in *Marine Geology* 56 (1984). Used by permission of Elsevier Science Publishers BV.

Figure 10.4. From C.D. Ford, 'Effects of glaciations upon karst aquifers in Canada' in *Journal of Hydrology* 61 (1983). Used by permission of Elsevier Science Publishers BV.

Figure 10.5. From C.D. Ford, 'The Physiography and Speleogenesis of Castleguard Cave, Columbia Icefields, Alberta, Canada' in *Arctic and Alpine Research* vol. 15, no. 4 (1983). Reproduced by permission of the Regents of the University of Colorado.

Preface

Canada has a variety of landscapes and landforms commensurate with its status as the world's second largest country. Nevertheless, it is grossly under-represented in the international geomorphological literature, which is dominated by Europe, the United States, and other more densely populated regions of the Earth. The general use of British or American textbooks in undergraduate programs has helped to ensure that Canada remains a geomorphological *terra incognita* for many students, continuing to propagate the view that the country is geomorphologically sterile or boring. There is of course nothing wrong, and indeed much that is right, in using examples drawn from around the world. Too often, however, the system is self-perpetuating, in that instructors, unaware of Canadian research in areas or topics outside their personal fields of interest, will refer to global examples that were introduced during their own undergraduate training. Canadian examples are therefore often ignored, even though in many cases they are superior to better-known examples elsewhere.

Canada is a land of geomorphological superlatives and peculiarities, which are often attributable to the legacy of glaciation. It supported the largest ice sheet in the northern hemisphere during the most recent ice age, and was the site of the first known ice age on Earth, more than two billion years ago. There are still extensive areas of both warm and cold ice in alpine glaciers and icefields, and in large Arctic ice caps. Half the country is underlain by permafrost, or perennially frozen ground, and virtually all of it was subjected to periglacial activity during the colder periods of the past. Canada's coast is the longest in the world, and it has more deltas, fiords, and estuaries than any other country. Factors such as the glacial steepening of slopes, deposition of sensitive marine clays, and the presence of frozen ground and, in places, high relief provide conditions suitable for a variety of large and small mass movements. Canada's drainage systems experienced enormous disruption during the last glacial stage. Its rivers and valleys still retain many vestiges of that period, although they are striving to adjust to the fairly recent onset of nonglacial conditions. Extensive areas of carbonate and sulphate rocks in Canada provide excellent examples of alpine and lowland karst.

One cannot, in researching a book of this nature, fail to be impressed by the contributions made by a fairly small group of workers in establishing the foundation on which Canadian geomorphological research is now based. Before the great growth in Canadian universities in the late 1960s and early 1970s, individuals often single-handedly studied huge areas of the country, or single major topics. Although there is now a much greater pool of active workers, the expansion of the geomorphological and geological communities has by no means been sufficient to eliminate exploratory or reconnaissance work, or the dominance of major fields of enquiry by one or two individuals.

This book is designed to be more in the nature of a systematic text than J. B. Bird's pioneering *The Natural Landscapes of Canada* (1972), and less of a regional survey. It is intended to provide a basis for junior undergraduates in geomorphology, and a continuing reference for senior undergraduates and postgraduates. With this aim in mind, a glossary is included to explain terms that may be unfamiliar to beginning students, while providing a text for the more experienced reader that is not burdened with too many unnecessary definitions. The fairly lengthy reference list at the back of the book will acquaint the reader with at least some of the more recent literature, and provide a basis with which to find more. Nevertheless, the scope of this book makes it impossible to acknowledge every source used in its writing. The author must therefore thank the numerous workers who have made this book possible, while apologizing for having in some cases omitted direct reference to their work.

I wish to thank Dr D.C. Ford, Dept. of Geography, McMaster University, and Dr D.M. Cruden, Dept. of Civil Engineering, University of Alberta, for reading and advising me on chapters 1, 8, 10 (D.C.F.), and 3 (D.M.C.). I am, however, entirely responsible for any errors or omissions in these and other chapters, including the accompanying figures.

ALAN S. TRENHAILE
UNIVERSITY OF WINDSOR

The Physical Background

The casual observer often has to travel great distances within Canada to notice major changes in scenery. This is particularly true of vast areas of the Canadian Shield in the stable interior of the country, and also of large parts of the Interior Plains that are arranged around it. Nevertheless, Canada has an enormous variety of landscapes within its borders. This is partly because of its size, which, with an area of almost 10,000,000 km^2, makes it the second largest country in the world. Changes in landscapes reflect differences in the surface rocks or **sediments**, and in the type and intensity of the climatically induced processes that have operated, or are still operating, on it. To describe and explain Canadian landscapes, therefore, we must have some basic understanding of the geology and climate of the country.

Plate Tectonics

The interior of our planet is still geologically active. It gives off heat and ejects volcanic material. It shifts during earthquakes, and generates horizontal and vertical movements of the crust. The interior of the Earth may be pictured as a series of concentric zones (Fig. 1.1). The central core is thought to consist of an alloy of iron and nickel, as well as some lighter elements, at a temperature of about 4,000°C. The inner portion of the core appears to be solid, but the outer portion is in a molten or liquid state. The mantle contains most of the Earth's volume and mass. Silica and oxygen predominate, with magnesium and iron as the most common metallic ions. The lower mantle is essentially solid, but there is a slow, steady flow of rock in the **asthenosphere**, a layer 100 to 400 km thick in the lower part of the upper mantle. Rock flow in the asthenosphere causes movements in

Figure 1.1.
*The interior of the Earth
(Stearn et al. 1979).*

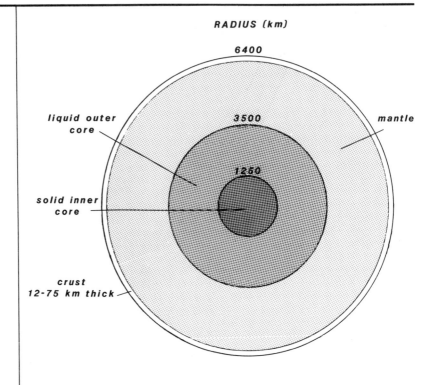

the overlying **lithosphere**. The surface skin, or crust, of the Earth is thickest under high mountains and thinnest under the ocean basins. The upper portions of the continental crust consist of **granitic** rocks, whereas the floors of the ocean basins are composed of **basaltic** rocks. Continental crust has more silica and potassium and less magnesium, iron, and calcium than oceanic crust, and it is therefore less dense.

The lithosphere, containing the crust and the upper part of the mantle above the asthenosphere, consists of six large plates (Eurasia, Americas, Pacific, Antarctic, Africa, and India-Australia), together with several smaller ones that fit in between. Most plates consist of oceanic and continental crustal segments, although the Pacific block is totally oceanic. While the boundaries between plates are the only zones that are able to generate earthquakes within the crust, deeper earthquakes can be generated at other points within the mantle.

There are three types of plate boundary:

(a) Oceanic ridges or mountain chains develop beneath the oceans along spreading centres, the boundaries of plates that are slowly moving apart. As the plates slide over the asthenosphere, separating at average rates of about 6 cm yr^{-1}, new crust is created from volcanic material welling up into the gap from the mantle. This material then moves outwards, away from the plate boundary, as the process continues.

(b) Deep oceanic depressions or trenches develop along subduction zones. These plate boundaries occur where one of a pair of converging plates is forced or deflected beneath the other into the asthenosphere, to be remelted and assimilated in the general circulation (Fig. 1.2). The distribution of most of the world's volcanoes and earthquakes is closely related to the occurrence of subduction zones. When oceanic and continental segments of the lithosphere converge, the denser oceanic plate is deflected beneath the continental plate. Nearly all oceanic lithosphere is eventually subducted into the mantle, but continental lithosphere is too thick and the continental crust too buoyant for subduction. Once it has been formed, rafts of continental lithosphere will therefore tend to continue to move over the surface of the Earth.

(c) The third type of plate boundary occurs where two plates slide past each other in opposite directions along a **transform fault**.

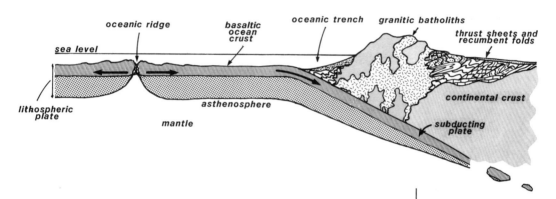

Figure 1.2.
Subduction, sea-floor spreading, and orogenesis (modified from Stearn et al. 1979 and Selby 1985).

The continents were joined together about 200 million years ago, forming a single supercontinent called **Pangaea**. The present distribution of land masses has been brought about by plate movement associated with the development of spreading centres and subduction zones, a process that continues today (Fig. 1.3).

Mountain Building

Mountains consist in part of sediments that accumulated in **depositional** basins along the margins of continents. Sediment may be laid down in trenches, on **continental shelves**, or on or at the foot of **continental slopes**. There are three main types of depositional basin:

(a) Basins at **passive** or inactive continental margins. Shallow-water marine sediments accumulate on the continental shelves, while wedges of deep-water sediment form at the foot of the continental slope and on the ocean floor. Slow subsidence of the lithosphere allows the sediment to accumulate to considerable thicknesses. These types of basin are characteristic of the Atlantic margins, including the area extending from Baffin

Figure 1.3.
Continental drift (Stearn et al. 1979).

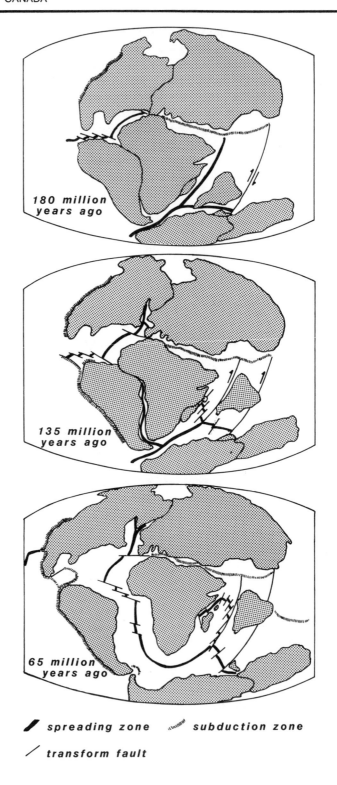

180 million years ago

135 million years ago

65 million years ago

▟ spreading zone   ⌇ subduction zone

╱ transform fault

Bay to Nova Scotia. The margin south of Nova Scotia was created by the separation of Africa and North America in the Jurassic period. Off Labrador and Baffin Island, the margin was formed by the separation of Greenland and North America in the late Cretaceous and Tertiary (Table 1.1).

(b) Basins at subduction boundaries. Large amounts of sediment are carried from the land across the continental shelves, and are swept by **turbidity currents** down the continental slope, along submarine canyons and into the trenches. There are four sedimentation areas: the trench, the fore-arc basin on the seaward side of the volcanic **arc**, the arc itself, and

Table 1.1 The geological timescale (based on Harland et al. 1982).

EONS	AGE	ERAS	PERIODS
P H A N E R O Z O I C	65	CENOZOIC	QUATERNARY[1] TERTIARY[2]
	248	MESOZOIC	CRETACEOUS JURASSIC TRIASSIC
	590	PALAEOZOIC	PERMIAN PENNSYLVANIAN[3] MISSISSIPPIAN[3] DEVONIAN SILURIAN ORDOVICIAN CAMBRIAN
P R E C A M B R I A N	2500	PROTEROZOIC	
	4000	ARCHEAN	
		PRISCOAN	

Ages are given in million of years before present.

[1]The Quaternary can be divided into the Pleistocene and Recent (Holocene) epochs.

[2]The Tertiary can be subdivided into the following epochs in order of decreasing age: Palaeocene, Eocene, Oligocene, Miocene, and Pliocene.

[3]The Mississippian and Pennsylvanian Periods are the North American equivalent of the European Carboniferous Period.

the foreland basin on the landward side of the arc. These basins are characteristic of the Pacific margins. Oceanic crust is subducted beneath continental crust in western Canada from the southern end of Vancouver Island to just south of the Queen Charlotte Islands, but the active margins are controlled by transform faults from the Queen Charlotte Islands to the southern Alaskan Panhandle.

(c) Basins between two masses of colliding continental crust.

Mountain-building or **orogenic** activity can result from the trapping and deformation of sediment and crust between two colliding continental masses, or massive **compressional** forces generated by continued subduction (Fig. 1.2). Orogenies do not occur as single events, however, but as the result of many episodes of compressional folding, **faulting**, overthrusting, uplift, **intrusion** of **batholithic** material, and vulcanism.

The Geological Evolution of Canada

Modern Canada is the product of three major geological developments (Stearn 1975, Stearn et al. 1979):

(a) the formation of the Shield;

(b) the formation of mountains from sediments that accumulated in basins around the margins of the Shield; and

(c) the deposition of sediments in shallow, or epeiric, seas in the intervening areas.

The Evolution of the Shield

In a number of recent publications, Paul Hoffman of the Geological Survey of Canada has shown that the ancient core of the continent was created between 1.98 and 1.83 billion years ago, through the violent collision and welding together of seven ancient microcontinents. These collisions resulted in extensive volcanic activity and deformation of the wedges of sedimentary rock that had been deposited on the sloping continental shelves. A large part of the Canadian Shield consists of Archean or early Precambrian rocks (Table 1.1), at least 2.5 billion years old. Most of the rock in the Shield is granite or **granite gneiss**, but there are also elongated belts of **greenstone** consisting of **metamorphosed sedimentary** (especially greywackes and conglomerates) and volcanic rocks (Fig. 1.4).

Proterozoic or late Precambrian volcanic and sedimentary rocks (particularly quartzose sandstones, limestones, and dolomites) lie **unconformably** on the granites. They are undisturbed in some places but folded and metamorphosed around granite intrusions in others. **Cuestas, mesas**, plateaux, flat plains, and other scenic elements that are generally untypical of the Shield have developed on Proterozoic sediments and basaltic lavas in several areas.

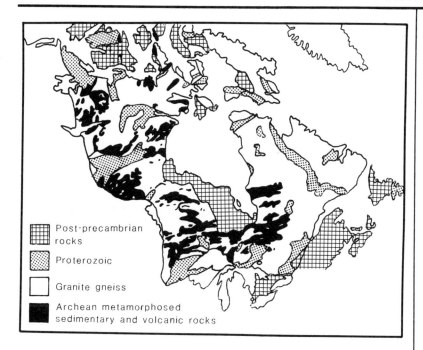

Figure 1.4.
The Canadian Shield
(Clark and Stearn 1960).

Legend:
- Post-precambrian rocks
- Proterozoic
- Granite gneiss
- Archean metamorphosed sedimentary and volcanic rocks

The Mountains

Mountains developed in three regions where enormous amounts of sediment accumulated in long, narrow basins, around the margins of the Canadian Shield (Fig. 1.5):

(a) the Cordilleran in the west;
(b) the Appalachian-Acadian in the east; and
(c) the Innuitian in the Arctic.

The Cordillera • The Cordillera has a very long and complex history, which is still only partly understood (Monger and Price 1979, Windley 1984, Gabrielse and Yorath 1989). It developed along a **tectonically active** continental margin, where an oceanic plate is subducted beneath a continental plate. It is a tectonic collage of **exotic** terrains that were swept eastwards by subduction and welded onto the western margin of the Americas plate. They include large blocks of oceanic crust, volcanic arc material, fragments of unknown continental margins, and oceanic plateaux. The large Stikine block in northern British Columbia, for example, was added to the western margin of North America in the late Triassic. The accretion of this large mass may have forced the subduction zone to the southwest and thrust older sediments to the east.

Shallow-water carbonate and quartzite sediments were deposited on the continental shelf and slope from the middle Proterozoic to the middle Devonian (1,500 to 380 million years ago). This formed a northeasterly tapering wedge (miogeocline) in what is now the eastern Cordillera. Oce-

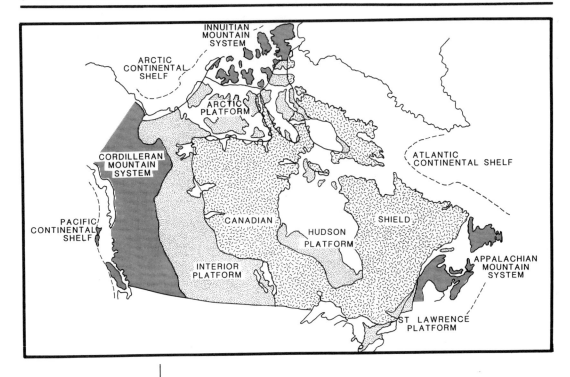

Figure 1.5.
The main structural units of Canada (Stearn 1975).

anic crust and volcanic arc deposits to the west were then thrust onto these wedge deposits in Mississippian and again in Permo-Triassic times. These low-angle thrusts carried enormous slabs of rock eastwards as they slid over each other. Exotic terrains, possibly including fragments of Asia, were accreted to the western parts of the basin through the late Palaeozoic and Mesozoic eras.

There was a major change in plate movements in the early Jurassic, when **Gondwanaland** started to break up as the Earth's spreading systems developed. The basic structure of the Cordillera was established from the middle Jurassic to the middle Cretaceous, between 165 and 95 million years ago. This was the result of compression in the eastern Cordillera, and **accretion** of exotic terrain and volcanic activity associated with subduction in the western Cordillera. The intrusion of molten material during this period formed batholiths that were eventually exposed at the surface by **erosion**.

Compression and **thrust faulting** in the eastern Cordillera and subduction-induced volcanic activity in the western Cordillera continued into the period between the late Cretaceous and the early Cenozoic (95 to 20 million years ago). Most of the Coast Mountains batholithic complex, which underlies the west-coast mountains for about 2,000 km, was emplaced during and just before this interval.

The formation of the Cordillera resulted from stresses generated by the subduction of the Pacific plate under the Americas plate. The two plates have interacted for almost one billion years, and both the recent elevation of the mountains and the occurrence of volcanoes and earthquakes show that this interaction continues today. The Coast Mountains of British Columbia, for example, have been raised by 2,000 to 3,000 m in the last ten million years. There were eruptions of large quantities of plateau basalt in southern British Columbia in the Oligocene and late Miocene epochs of the Tertiary Period. At least five volcanoes in the Stikine Ranges erupted beneath glacial ice (Allen et al. 1982), and postglacial vulcanism occurred in the Anahim and Stikine belts of central and northern British Columbia, respectively. There are also small **pyroclastic** domes and lava flows of **Holocene** age in the Coast Mountains, the most recent eruption having occurred less than 150 years ago. This area has also experienced several large earthquakes in historical times, including one on Vancouver Island in 1946.

The Appalachian-Acadian Region • The rocks of the Appalachian-Acadian Region range from Precambrian to early Mesozoic (Windley 1984). The development of the Canadian Appalachians is thought to reflect the opening and closing of the Iapetus Ocean, the predecessor of the North Atlantic. The first sediments were deposited as the ocean began to open up in the late Precambrian. Material was eroded from nearby Precambrian mountains, and deposited in wedges on the continental shelf. Sedimentation continued as the ocean expanded during the Cambrian and early Ordovician. A shallow-water, carbonate-rich bank of sediment developed on the continental shelf, with deep-water **clastic** material on the **continental rise** to the east. There was also volcanic activity associated with an island arc, while sediments and volcanics were scraped from the ocean floor by subduction and deposited in the active trench.

The Taconic Orogeny in the middle to late Ordovician may be attributed to the closure of the Iapetus Ocean and the destruction of its continental margin. Block faulting resulted in the fragmentation and partial subsidence of the carbonate shelf, and in northwest thrusting, folding, and metamorphism. The Acadian Orogeny, which occurred at the end of the lower Devonian, about 380 million years ago, is more difficult to explain. Since there is no evidence for the existence of an ocean or continental margins at that time, the cause of this episode cannot be traced to subduction and moving plates. Rather, the Acadian Orogeny has been attributed to deformation of **successor basins** that developed across the destroyed margins and oceanic tract of the Iapetus Ocean (Williams 1979). The orogeny was followed by uplift and erosion, resulting in the deposition of enormous amounts of coarse sediment from the middle Devonian to the Permian. The Pennsylvanian or lower Permian Alleghanian Orogeny, resulting from the collision of North America and Africa, affected only the southern and central Appalachians of the United States.

Mountain-building had therefore welded the Appalachian-Acadian sedimentary basin onto the southeastern rim of the Shield by the end of the

Palaeozoic era. The Appalachian Mountains were probably then comparable in size and height to the Cordillera today, but they are now an old chain and have been substantially lowered by erosion.

The Innuitian Region • A mountain system similar to the Appalachians can be traced across the High Arctic from Alaska to the northern coast of Greenland (Fig. 1.6) (Trettin and Balkwill 1979). There was rapid deposition and continuous subsidence in a southeasterly trending basin (the Franklinian) from Cambrian to Devonian time. A major source of these sediments was provided by intermittently rising mountainous terrain to the northwest. Although this area is largely offshore today, at times it expanded into northern Ellesmere Island. Three major depositional belts developed in the marine basin. A rapidly subsiding shelf in the southeast received largely carbonate sediment. The Hazen Trough in the central portions of the basin developed as a result of the gradual deepening of a shelf basin. The existence of this deep trough has been identified on northern Ellesmere Island, and it has been tentatively extended to Melville Island. A complex area in the northwestern part of the basin contained a volcanic island arc, shelf, and coastal plain, which are reflected today in the presence of vol-

Figure 1.6.
The Arctic Islands (modified from Stearn et al. 1979).

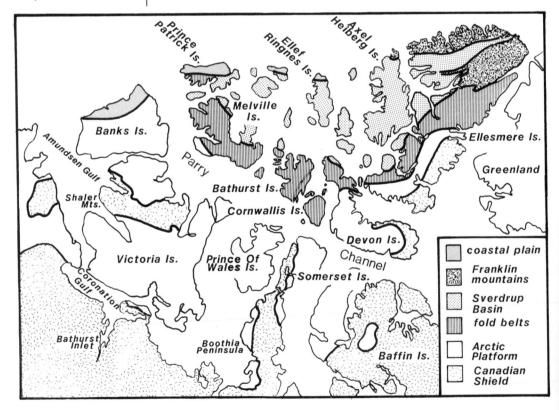

canic rocks, carbonate sediments, and coarse, shallow marine and non-marine sediments.

The greatest deformation of the Innuitian Region occurred during the Ellesmerian Orogeny in Devonian-Pennsylvanian time. Deformation progressed from the north to the south, forming three belts with different structural characteristics. There was folding, faulting, uplift, and granitic intrusion in northern Ellesmere and Axel Heiberg Islands, folding and faulting in the northeastern portion of the central Hazen Trough, and **skin-type** deformation in the central Ellesmere fold belt. This orogeny may have been associated with the presence of a Palaeozoic ocean north of Ellesmere Island and Greenland, which was probably connected to the proto-Atlantic Iapetus Ocean. The Mississippian to late Cretaceous interval was dominated by the subsidence and filling of the Sverdrup Basin.

The Epeiric Seas

Chemical and organic sediments accumulated in the shallow, or epeiric, seas that extended over the stable interior of the continent. Deposition occurred over large areas of the Shield, particularly in the upper Ordovician and the middle Silurian. Areas flooded during marine **transgressions** stood as low hills and coastal plains during periods of lower sea level. The sediments deposited in these shallow waters were much thinner than those in the marginal marine basins, and they consist of cleaner rocks, including limestones, dolomites, and sandstones. The rocks are commonly faulted, but folding is rare because of the enormous strength of the underlying Shield. The last major continental transgression took place in the late Cretaceous in the western Interior Plains and the western District of Mackenzie. This resulted in the deposition of great thicknesses of shale, clay, and sandstone, which now underlie glacial deposits over large areas. Other large areas of Palaeozoic rock include the Hudson Bay Lowlands, the southern Arctic Islands, southern Ontario, and the St Lawrence Lowlands. Erosion has reduced much of the formerly extensive Palaeozoic cover on the Shield, however, to small but widespread patches, which are usually in **grabens** or other depressions.

Physiographic Regions

Canada's three main structural components account for many of its geological and physiographical characteristics (Fig. 1.5) (Bostock 1970). The ancient Precambrian crystalline rocks of the Shield occupy nearly half the country. This surface can be compared to an inverted military shield descending outwards from a flat, slightly depressed centre, which is occupied by Hudson Bay and a basin containing about 3,000 m of younger sediments. Palaeozoic rocks may once have extended over the whole of the Shield, but within the area in which the Shield is now exposed, they remain only on the Hudson Platform and in the Foxe Basin, and as several small, scattered erosional remnants and down-faulted blocks.

The Borderlands surrounding the Shield are composed of segments of two concentric rings of mainly sedimentary rock. The inner ring consists of lowlands and plains, while the outer ring is made up of mountain chains. The two Borderland rings can be subdivided into regions or provinces, based on more distinct changes in landscape and geology, more easily than is possible on the Shield.

The Physiography of the Shield

Many geologists believe that the Canadian Shield was reduced to an almost featureless surface during the Archean era, then covered beneath Proterozoic and Palaeozoic sediments. Erosion later exhumed this ancient surface, which, because of the resistant nature of the rocks, has since been preserved with only minor modification.

Although there are considerable differences in climate, and therefore in the **geomorphic** processes operating on the Shield, the dominance of gneissic rocks has produced a general uniformity in the landscapes of this vast region. The Shield is therefore usually subdivided using structural rather than topographic criteria. Seven structural provinces and several subprovinces have been defined, based upon differences in the structural trends and the style of folding within the Shield (Fig. 1.7). Each structural province contains rocks of many ages, which have been subjected to more than one period of orogenic activity. Within each province, however, most of the metamorphic rocks are of about the same age, and the effects of one major orogeny are dominant. The Slave and Superior Provinces, for example, were metamorphosed about 2,500 million years ago in the Kenoran Orogeny, whereas the Nain and Grenville Provinces were last deformed 1,300 and 1,000 million years ago, respectively.

Most of the Shield is rolling or undulating, with a regular skyline broken by rounded or flat-topped hills, and a local **relief** of between 60 and 90 m. More than a quarter of the land in many areas is covered by water, in swamps, ponds, and lakes. Often, therefore, the only scenic variety is provided by differences within the approximately 20 per cent of the Shield that is not underlain by granites or granite gneisses.

The few types of landscape that differ from the normally rolling scenery of the Shield include (Fig. 1.7):

(a) Plains developed in horizontal or gently **dipping** Proterozoic sedimentary rocks, including sandstones, limestones, conglomerates, and dolerites. They include the Thelon Plain in the Northwest Territories, the Athabasca Plain in northern Saskatchewan, and the smaller Cobalt Plain in northern Ontario. The Nipigon Plain in northwestern Ontario is underlain by flat Proterozoic sediments and volcanic **sills**.

(b) Hills found in several areas in Proterozoic volcanics and folded or dipping sedimentary rocks. Large cuestas have formed on **gabbro** sills or basaltic flows in, for example, the Coronation Gulf–Bathurst Inlet area of the Northwest Territories, and around Thunder Bay in Ontario.

(c) Mountains and uplands in several areas. The Laurentian Highlands rise from the north shore of the St Lawrence Lowlands to maximum elevations of more than 1,200 m, although most of the area is at about 600 m. The Mealy Mountains of southeastern Labrador consist of particularly resistant rocks that rise abruptly from their hinterland to a maximum height of 1,130 m. The Labrador Highlands and the Davis Highlands of Baffin, Devon, and Ellesmere Islands consist of deeply dissected crystalline or sedimentary rocks incised by fiords, deep glacial valleys, and cirques. The Torngat Mountains in the Labrador Highlands, which are the highest on the mainland of eastern Canada, and the peaks of the Davis Highlands rise to more than 1,500 m above sea level.

Faulting defines parts of the Shield boundary, particularly in the southeast, and it has brought different types of rock into contact with each other within the Shield. Valley-like depressions, or grabens, are created when a block subsides between two parallel faults. There are many examples on

Figure 1.7.
The structural provinces of the Canadian Shield (Stockwell 1982).

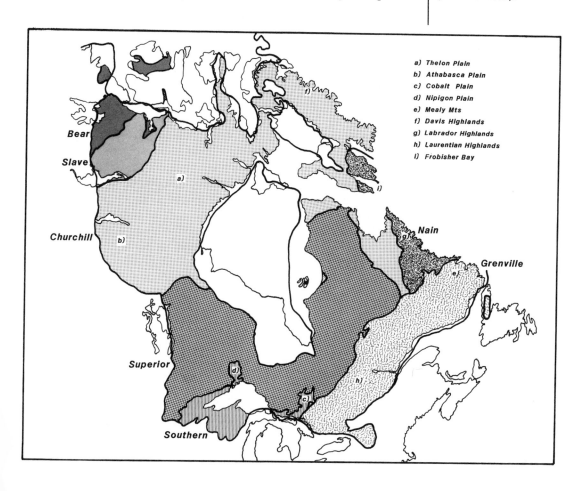

a) Thelon Plain
b) Athabasca Plain
c) Cobalt Plain
d) Nipigon Plain
e) Mealy Mts
f) Davis Highlands
g) Labrador Highlands
h) Laurentian Highlands
i) Frobisher Bay

Bear
Slave
Churchill
Superior
Southern
Nain
Grenville

the Canadian Shield, including Frobisher Bay on Baffin Island, Bathurst Inlet in the Northwest Territories, and the Ottawa Valley above Ottawa. Granitic surfaces on the Shield are crossed by linear depressions etched along joint and fracture systems. These depressions could have been excavated by frost action, or by chemical weathering in the warm Tertiary period, with streams or glaciers removing the debris. Modification of the Shield by ice was probably greatest along its northeastern rim, where deep fiords were cut by outlet glaciers flowing from the highlands. Elsewhere, the rocks of the Shield have been smoothed and abraded, but many of the lakes are the result of glacial deposits damming linear depressions, rather than of deep glacial erosion of rock basins.

The Lowlands and Plains

A great series of plains forms a wide crescent around the northwestern, western, and southern perimeter of the Canadian Shield. These plains have formed on a wedge of outwardly thickening, generally flat-lying and undisturbed, sedimentary rocks, which were deposited in shallow (epeiric) seas onto the underlying Shield. This inner zone includes much of the southern Arctic Islands, areas on either side of the Mackenzie River, the Prairies, the Great Plains, the Midwest and adjacent regions of the United States, and the Great Lakes–St Lawrence Lowlands of Canada.

The Arctic Lowlands • The Lowlands, which include most of the Arctic Islands south of the Parry Channel, are formed on Palaeozoic and late Proterozoic sedimentary rocks between the Innuitian Region to the north and the Shield to the south (Fig. 1.6). The Lancaster Plateau slopes gently southwards from southern Ellesmere Island across Devon, Somerset, and northwestern Baffin Islands into the Boothia Plain. The Foxe Plain is a shallow basin area underlain by Palaeozoic rocks in southwestern Baffin Island. The surface of the Victoria Lowlands is generally smooth and undulating, with extensive glacial deposits. The Shaler Mountains on Victoria Island consist of a dome of Proterozoic sedimentary rocks intruded by gabbro sills.

The Northern Interior Plains • The Plains begin at Amundsen Gulf, a part of the Arctic Slope that drains directly into the Arctic Ocean. The rolling, low-relief terrain, extending from the Arctic coast to south of Great Slave Lake, is generally underlain by Palaeozoic or Mesozoic bedrock, or glacial deposits. The Alberta Plateau to the south consists of a ring of plateaux with very wide valleys. More than half this area is occupied by the Peace River and Fort Nelson Lowlands.

The Western Interior Plains–the Prairies • The southern Prairies are composed of three main surfaces, increasing in elevation to the west (Fig. 1.8). The Manitoba Plain (the First Prairie Level) and its extension to the north

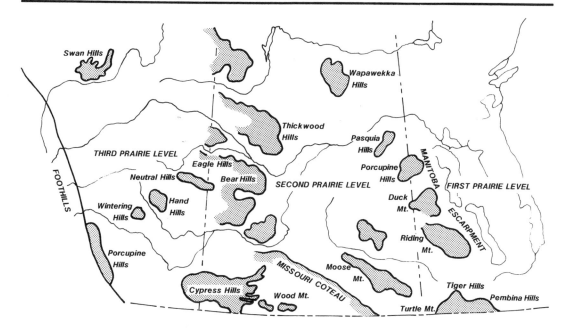

Figure 1.8.
The western Interior Plains (Prairies).

of Great Slave Lake are essentially level, glacially modified rock surfaces with some low plateaux, largely consisting of Palaeozoic limestones and dolomites. This surface terminates in the west at the Manitoba Escarpment, which is dissected by valleys into a number of uplands. The escarpment is formed in Cretaceous shales and other sedimentary rocks that underlie much of the western and central Prairies. The scenery of the Saskatchewan Plain (the Second Prairie Level) is dominated by the effects of glaciation, although some slightly higher areas correspond to low domes in the Cretaceous rocks below. The Saskatchewan Plain terminates in the west at the Missouri Coteau, a low escarpment in upper Cretaceous shales and clays marking the eastern margin of the Albertan Plain (the Third Prairie Level). The landscapes of the Albertan and Saskatchewan Plains are broadly similar. Hills and plateaux, including the Wood Mountain Plateau and the Cypress Hills, consist of Tertiary rocks, which have resisted erosion more successfully than their surroundings.

The Great Lakes–St Lawrence Lowlands • This region is a discontinuous strip extending from the Strait of Belle Isle to the western shore of Lake Huron. The Canadian Shield in eastern Ontario separates the region into an eastern and a western section. In southern Ontario, a series of Palaeozoic limestones, dolomites, and shales was deposited onto a foundation of Shield rock. The Niagara Escarpment is the most prominent and best-known example of several escarpments in the resistant, southwesterly-dipping rocks. Glacial deposits cover most of southern Ontario, however, and the details of its scenery are largely the product of the last glacial stage. The St Lawrence Lowlands extend from west of Ottawa to just east of Québec

City. The area is low-lying, except for the Monteregian Hills, a series of **igneous** intrusions in the Montreal Plain. The submergence of this area by the Champlain Sea left very flat, poorly drained plains (Chapter 5; see also fig. 3.5). An extension of the lowlands to the east includes Anticosti Island, the Mingan Islands, a few small areas along the Gulf of St Lawrence and the Strait of Belle Isle, and parts of northwestern and southwestern Newfoundland.

The Mountains

The outer ring of the borderlands consists of three mountainous regions composed of intensely folded and deformed rocks.

The Cordillera • Along most of its length, the Cordilleran Region consists of a longitudinal series of plateaux and plains with mountains and highlands on either side. The interior plateaux are absent, however, in part of northern British Columbia, where the mountains extend from the Pacific to the northern Rockies. The tripartite division re-establishes itself further north in northern British Columbia and the Yukon, although in a more complex arrangement than to the south. A series of east-west trending plateaux, including the northern part of the Interior Plateau of British Columbia, and the Liard, Yukon, and Porcupine Plateaux of the Yukon, also divide the Cordillera into transverse sections (Fig. 1.9).

The Eastern System of the Cordillera consists of folded and thrust-faulted sedimentary rocks. The mountains in the Rockies of the southern Cordillera are commonly more than 3,000 m in height, attaining their maximum of 3,962 m in Mount Robson. The Rocky Mountain Trench and two similar trenches in the Yukon are very long, straight, and wide valleys. These trenches could be the result of faulting, or they could represent the upper portions of thrust sheets, although they were extended and modified by stream and, to a lesser extent, glacial action. The Rockies are replaced in the northern Cordillera by a series of lower and less continuous ranges, with their associated plateaux, plains, and lowlands.

The Interior System of the Cordillera contains folded sedimentary, volcanic, and metamorphic rocks, and igneous intrusions. Consisting of a number of intermontane plateaux, it also includes the Columbia and Cassiar-Omineca Mountains of British Columbia.

The Western System itself consists of three longitudinal systems – the Coast Mountains in the east; the mountains of the Queen Charlotte Islands and Vancouver Island in the west; and the largely submerged Coastal Trough – with small lowland areas between. The Coast Mountains diminish in width to the north, but they increase in height, culminating in Mount Logan at 6,050 m above sea level.

The Appalachians • The Appalachians are the remnants of mountain ranges formed during the Palaeozoic. The erosion of different rock types and structures has produced a variety of landscapes, including lowlands,

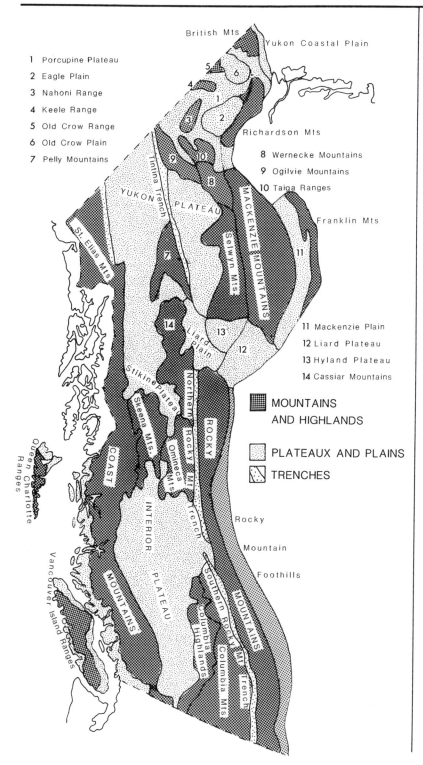

Figure 1.9.
The physiographic regions of the Cordillera (Bostock 1970).

1 Porcupine Plateau
2 Eagle Plain
3 Nahoni Range
4 Keele Range
5 Old Crow Range
6 Old Crow Plain
7 Pelly Mountains

8 Wernecke Mountains
9 Ogilvie Mountains
10 Taiga Ranges

11 Mackenzie Plain
12 Liard Plateau
13 Hyland Plateau
14 Cassiar Mountains

MOUNTAINS AND HIGHLANDS

PLATEAUX AND PLAINS

TRENCHES

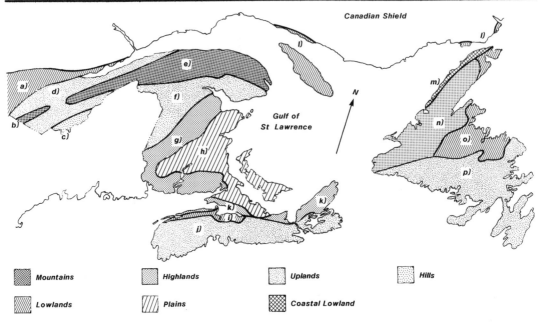

a) Central St Lawrence Lowland   b) Sutton Mts   c) Megantic Hills   d) Eastern Quebec Upland   e) Notre Dame Mts
f) Chaleur Uplands   g) New Brunswick Highlands   h) Maritime Plain   i) Annapolis Lowland   j) Atlantic Uplands of Nova Scotia
k) Nova Scotia Highlands   l) East St Lawrence Lowland   m) Newfoundland Coastal Lowland   n) Newfoundland Highlands
o) Newfoundland Central Lowland   p) Atlantic Uplands of Newfoundland

Figure 1.10.
The physiographic regions of eastern Canada (Bostock 1970).

uplands, and highlands. The region can therefore be divided into numerous physiographic subdivisions (Fig. 1.10). Three broad regions can be distinguished on Newfoundland. The Palaeozoic and Precambrian highlands in the west are areas of rugged terrain, ranging up to 800 m in height. The landscape is more rolling in the uplands of southern Newfoundland, and relief is normally quite low. The lowlands of central Newfoundland are gently rolling, and largely covered by glacial deposits. Uplands occupy most of eastern Nova Scotia, as well as a narrow basaltic strip along the southern shore of the Bay of Fundy. A highland region extends from the Cobequid Mountains in the west to the Cape Breton Highlands in the east. The main lowlands are in the Annapolis Valley, along the southern shore of the Minas Basin, and in northern Nova Scotia, continuing northwards into eastern New Brunswick and Prince Edward Island. The highlands of New Brunswick are distributed in a U-shape in the central and southern parts of the province, and are separated from the Notre Dame Mountains to the north by the Chaleur Uplands. The Notre Dame Mountains attain elevations of more than 1,200 m on the Gaspé Peninsula, but they become lower and merge with the uplands of eastern Québec to the west.

The Innuitian Region • The Innuitian Region is a roughly triangular area in the High Arctic, on the islands to the north of the Parry Channel (Fig.

1.6). The mountains on Axel Heiberg and northwestern Ellesmere Islands, in the northern part of the region, are essentially long ridges of folded Palaeozoic and Mesozoic rocks. They attain elevations of up to 2,400 m above sea level, but they are almost buried beneath ice sheets in some areas. Folded Palaeozoic rocks have also produced an extensive belt of rugged scenery along the whole of the southern part of the region; with the exception of northeastern Ellesmere Island, however, this area tends to be considerably less mountainous than further north. Landscapes in the fairly weak Mesozoic and Tertiary sediments between the two rugged belts are generally more subdued, consisting of rolling or undulating uplands and plateaux.

On the basis of the crude physiographic symmetry of the continent of North America, the narrow Arctic Coastal Plain can be considered to be the northern counterpart of the much wider coastal plain of the southern and southeastern United States. The Arctic Coastal Plain extends along the northwestern extremities of mainland Canada and the Arctic Islands (Fig. 1.6). It is underlain on the Arctic Islands by Tertiary or early Pleistocene sand and gravels, and varies, according to the degree of uplift and dissection, from a low, flat surface to rolling, hilly terrain. The Coastal Plain on the mainland includes the Mackenzie Delta and the Yukon Coastal Plain.

Meteorite Craters

It has been accepted for a long time that the Earth's surface is shaped by erosional, tectonic, and volcanic processes. Increasingly, however, a fourth factor, impact cratering, is also being considered. Meteorite impact craters (astroblemes) are often difficult to recognize because of the fairly youthful surface of the Earth, and the masking effects of geomorphic processes. We have no record of the violent bombardment of meteorites in the early history of our solar system, and erosion and infilling have obscured or probably even obliterated the effects of more recent impacts. A large proportion of the meteorites colliding with the Earth must have fallen into the oceans, which constitute about 70 per cent of its surface. Yet we lack all evidence of submarine impact craters, except for one recently identified about 200 km southeast of Nova Scotia. This crater, which is about 45 km in diameter, was formed in the early Eocene.

Most craters have been found on the old, stable shields of the continents, and on the surrounding undisturbed sedimentary rocks. Twenty-four impact craters have been positively identified in Canada (Fig. 1.11) (Grieve and Robertson 1987, MacLennan 1988), although further investigation may confirm the existence of others among the hundreds of approximately circular topographic features that have been recognized. The frequency of cratering has decreased through time, and it is partly for this reason that most Canadian craters are on the Shield. They range in size from the Manicouagan Crater, with a diameter of about 100 km, down to a number of smaller forms with diameters of less than 10 km. Their ages range from

Figure 1.11.
*Impact craters in Canada
(Grieve and Robertson
1987).*

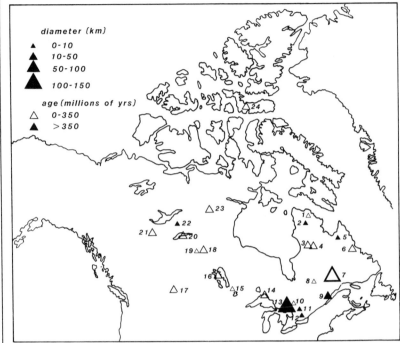

1. New Quebec, Que. 2. Lac Couture, Que. 3. Clearwater L. West, Que. 4. Clearwater L. East, Que. 5. Lac La Moinerie, Que. 6. Mintastin, Nfld. and Lab. 7. Manicouagan, Que. 8. Ile Rouleau, Que. 9. Charlevoix, Que. 10. Wanapitei L., Ont. 11. Brent, Ont. 12. Holleford, Ont. 13. Sudbury, Ont. 14. Slate Is., Ont. 15. West Hawk L., Man. 16. Saint Martin, Man. 17. Eagle Butte, Alta. 18. Deep Bay, Sask. 19. Gow L. Sask. 20. Carswell, Sask. 21. Steen River, Alta. 22. Pilot L., NWT. 23. Nicholson L., NWT. 24. Haughton, NWT.

5 million years in the case of the New Québec Crater to 550 and 450 million years for the Holleford and Brent Craters, respectively. Many workers now believe that the Sudbury Basin, which has a diameter of roughly 140 km, was formed by meteoric impact about 1,850 million years ago.

Where erosional processes are negligible – for instance, on the moon – two main types of crater can be distinguished: simple or bowl-shaped forms with a raised bedrock rim, about one-third as deep as they are wide; and larger (diameters usually greater than 4 to 5 km), more complex forms that are often no more than about one-tenth as deep as they are wide. These large, shallow craters have little or no rim, but they have a prominent uplifted central area. On Earth, erosion and sedimentation have greatly reduced the depth of meteorite craters. The large Manicouagan Crater consists of two crescent-shaped lakes (now coalesced as a result of the building of dams) around a central area more than 500 m higher than its surroundings. Probably because of infilling with sediment, this structure is now far shallower, relative to its width, than features of similar diameter on the moon.

Earthquakes

Each year, between about 200 and 300 earthquakes in Canada are of sufficient magnitude to be recorded. Of these, 59 per cent occur in the north,

27 per cent in the west, and 14 per cent in the east (Whitham 1975). Most are quite weak, but there have been several powerful tremors in historical times. Earthquakes have been particularly numerous in the St Lawrence Valley, around Vancouver and off the coast of southern British Columbia, in the Yukon and along the Mackenzie Valley, and on eastern Baffin Island and elsewhere in the eastern and central Arctic. Those off the coast of British Columbia appear to be associated with plate boundaries, but it is difficult, on the basis of plate tectonics, to account for the large numbers that occur on southern Vancouver Island and in Georgia Strait and Puget Sound. Most of the earthquakes in the Arctic and eastern Canada are not associated with plate boundaries. Earthquakes in the Baffin Island and Boothia–Ungava areas may be generated by **isostatic** rebound. Glacially-induced uplift may also trigger earthquakes in other parts of the Arctic and in eastern Canada, although they are primarily the result of stresses associated with rift faults and other structural phenomena (Quinlan 1984, Adams and Basham 1989). On 25 November 1988, for example, an unusually strong intraplate earthquake in the Saguenay region of Québec registered 6 on the Richter scale. The epicentre was estimated to be at a depth of 29 km below the surface, near the lower edge of the Earth's crust. The typically deep earthquakes that occur in this region are thought to be associated with the presence of deep faults.

Climate

Landscapes develop through time as climatic processes operate on the Earth-forming materials. The diversity of Canadian landscapes is therefore the result of climatic as well as geological variations. Differences in climate reflect the influence of latitude, distance from the sea, the presence of mountain barriers, and the degree to which areas can be invaded from outside by warm and cold air (Hare and Thomas 1974).

Temperature is largely determined by the latitude, elevation, distribution of land and water, and prevailing winds and storm tracks (Fig. 1.12). Although cooled during winter, the Pacific Ocean retains a huge amount of heat, which warms the winds blowing over it. The mild winters of British Columbia are the result of westerly winds sweeping onto the land from the Pacific, but because the ocean heats up slowly, the same winds cool the land in summer. Far away from the sea, the western Interior Plains and most of northern Ontario experience a continental climate, with warm to hot summers and very cold winters. Mean January temperatures are frequently about -15°C or less. The temperature can fall to below -40°C with the influx of Arctic air from the north, but it may be only just below freezing when milder Pacific air enters the region. Winter temperatures are higher in the Chinook belt of the western Interior Plains, when cold Arctic air is displaced by warmer Pacific air descending off the Rockies. This can cause temperatures to rise from -35°C to +5°C in only a few hours. Summers are warm in southern Ontario, and the influx of warmer air from the southwest provides milder winter conditions than in the Prairies. Despite

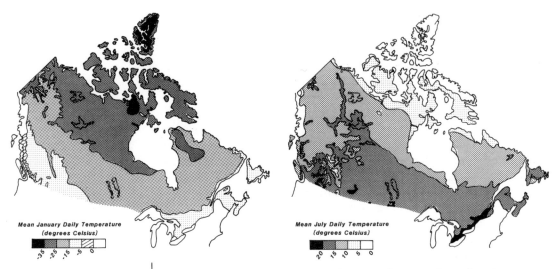

Mean January Daily Temperature
(degrees Celsius)

Mean July Daily Temperature
(degrees Celsius)

Figure 1.12.
*Mean January and July
daily temperature* (Hydrological Atlas of Canada
1978).

its maritime location, the preponderance of eastward-moving storms over Atlantic Canada produces a continental type of climate. These westerly air masses provide much greater ranges of temperature from day to day and from season to season than could occur in a true maritime climatic region. Nevertheless, the area is subject to the usual modifying effect of the sea. The coasts of Newfoundland and Nova Scotia are cooled in summer by the cold Labrador Current, and there is frequently **pack ice** along the eastern coast of Newfoundland in winter. Spring arrives later in the Atlantic Provinces than in Ontario because of the cooling effect of northerly and easterly airstreams passing over the cold, ice-strewn water.

The Arctic climatic region is an area of treeless, open **tundra** and permanently frozen ground north of the treeline. The Boreal climatic zone is the climate of the Boreal forest; a forest of conifers and some hardy broad-leaved trees distributed around the southerly perimeter of the Arctic tundra. The Arctic and Boreal regions are extremely cold, with mean annual temperatures near or below freezing. The ground cools in the fall and winter because of the lack of solar radiation. Cooling stops in the spring with the return of solar radiation and the invasion of warmer air, although up to 80 per cent of the solar radiation may be lost by reflection from the snow cover. Temperatures quickly increase once the snow finally melts, but cooling soon predominates again in the fall, as the noonday sun becomes lower in the sky and the length of the nights increases. Another major reason for the severity of Arctic climates is the dominance of cold airstreams from the pack ice of the Arctic Ocean and the Greenland ice cap.

Arctic airstreams also dominate Boreal climates in winter and spring, but westerly airstreams from the Pacific are more common in the summer and fall. Arctic and Boreal climates extend furthest south in eastern Canada, because the change from Arctic to westerly dominance occurs earlier in the west than in the east, and because greater snowfall in the east reduces the warming effect of spring sunshine.

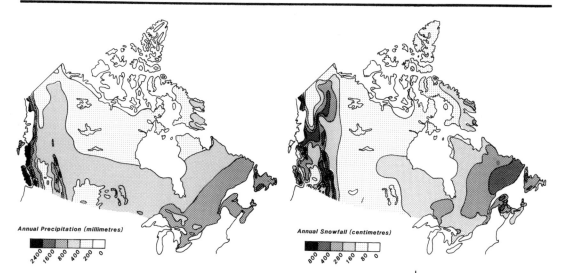

Annual Precipitation (millimetres)

2400 1600 800 400 200 0

Annual Snowfall (centimetres)

800 400 280 160 80 0

Despite the severe climate, solar radiation and the influence of warmer air moving off the land are sufficient to cause the eventual breakup of winter ice in Hudson Bay, Foxe Basin, Baffin Bay, and the coastal waters between the Arctic islands. It does not completely disperse in Hudson Bay until late August or early September, however, and can persist throughout the summer in the channels leading to the Arctic Ocean. There is permanent ice cover on the Arctic Ocean itself, except for some open patches or channels that can exist in all seasons.

There are two main areas of fairly high precipitation in Canada (Fig. 1.13). Most of the upland areas of the Cordillera are well watered, particularly on the western-facing slopes. Annual mean precipitation can approach 5,000 mm on western Vancouver Island, and it is almost as great on the Coast Mountains on the mainland. Snowfall is very heavy throughout the Cordillera, especially on the western sides of the ranges, although precipitation declines very rapidly to the east of the Rocky Mountains. The second wet area is in eastern Canada, east of a line running from southern Baffin Island to Winnipeg. This area is somewhat drier than the Cordillera, although annual precipitation, with a heavy snowfall component, is more than 1,250 mm in the Laurentides and in the hillier areas of the Maritime provinces. Most moisture in the Cordillera is derived from westerly winds blowing in from the Pacific, whereas in the east, the main sources are the Atlantic Ocean and the Gulf of Mexico. The western Interior Plains, the deep valleys of the interior of British Columbia in the rain shadow of the Coast Mountains, and the Arctic and western Boreal regions are much drier areas. Annual precipitation may be less than 125 mm in the High Arctic Islands, and it is usually less than 500 mm throughout the Arctic and Boreal regions. Nevertheless, because evaporation is low in the north, low precipitation does not produce aridity, whereas the combination of low precipitation and high evaporation does generate aridity and dust-bowl conditions in southern Alberta and Saskatchewan.

Figure 1.13.
*Annual precipitation and snowfall (*Hydrological Atlas of Canada 1978).

2 | Weathering

Weathering is essentially the *in situ* breakdown and alteration of Earth-forming materials. Many rocks were formed within the crust of the Earth under conditions of high temperature and pressure, and in the absence of air and water. Weathering may therefore be viewed as the response of material now at or near the Earth's surface to lower pressures and temperatures, and to the presence of air, water, and biological organisms. Weathering reduces the strength of rock and increases its **permeability**, thereby helping other erosive agents, which remove and transport the weathered products by mechanical and solutional processes.

Physical or Mechanical Weathering

Physical weathering involves the mechanical breakdown of rocks into progressively smaller units. This is caused by physical forces generated from within or outside the rock by a number of mechanisms.

Pressure Release

The removal of surface material by **erosion** reduces the confining pressure on the underlying rock. This allows the mineral grains to move further apart, creating larger voids in the rock. Expansion or dilation of the rock perpendicular to a valley side, cliff, or some similar erosional surface divides the rock into sheets, along cracks that are parallel to that surface. These cracks can be large **joints** or small and shallow fractures. Pressure-release, or dilation, joints are most common in granites, but are also found in a variety of other rocks. Explosive rockbursts on the floors of mine passageways cut in granite or other dense rocks can occur with fatal consequences. The development of dilation joints causes rock falls and other kinds of mass movement, and encourages further recession of slopes cut by glaciers, waves, rivers, man, and other erosional agencies (Fig. 2.1). In the Coast Mountains of British Columbia, for example, such joints may control the shape of valley sides and cirque headwalls and influence the nature of cirque development (Ryder 1981). Pressure release also assists the work of other weathering processes by opening up fissures in the rock.

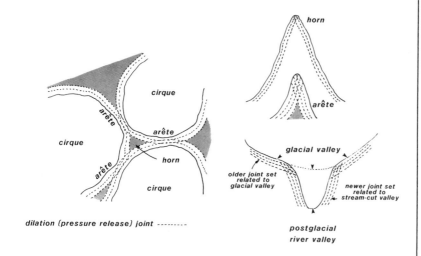

Figure 2.1.
Some glacial examples of pressure release and the formation of dilation joints.

Frost Action

It is generally assumed that rocks in cold regions are split or shattered by the alternate freezing and thawing of water contained in crevices, or in small voids and capillaries (gelifraction). There is considerable debate, however, about the precise mechanisms involved in frost weathering. Several mechanisms have been suggested, including the pressures generated in rock crevices by freezing water; the growth of ice crystals in rock capillaries; the pressure of water driven from a freezing area by the growth of ice crystals; osmotic pressures; and the freezing of water expelled into adjacent crevices by unfreezable ordered water in the fine capillaries of the rock (see Wetting and Drying, p. 27).

Water expands by about 9 per cent on freezing. This can generate very high pressures, up to 214 MPa at -22° (the maximum of 215 MPa is reached at -40°C). But these high pressures could be attained only under ideal conditions, which are unlikely to exist in the field. Actual pressures are therefore probably much less than the maximum.

Workers have been particularly interested in determining the climatic conditions that are most conducive to frost action. Effective frost action can occur only in environments with a plentiful supply of water and suitable fluctuations in temperature. Many sites appear to lack at least one of the essential requirements. Most researchers have found that in the laboratory, as long as there is enough water to saturate the rock samples, the damage increases with the frequency of the freeze-thaw cycles, and consequently with the rate of freezing, rather than its intensity. Nevertheless, the evidence is often contradictory. On the Niagara Escarpment on Bruce Peninsula in southern Ontario, the amount of debris in frost-induced rock falls increases with the length of the freezing period, rather than with the intensity of freezing, or the frequency of the frost cycles. As expected, however, most debris was produced when there was enough water for the rocks to attain fairly high levels of saturation (Fahey and LeFebure 1988).

Salt Weathering

Salt weathering is probably most important in coastal and arid areas. Three main mechanical processes are involved:

(a) Salt crystals in the small capillaries of rocks can generate destructive pressures as they absorb water. The greatest pressures are exerted at low temperatures with high relative humidity. The alternate hydration and dehydration of salt crystals contained within rocks can occur several times in one day, particularly where there are frequent changes in temperature and humidity. Carbonate and sulphate salts expand particularly rapidly and by significant amounts when hydrated, although sodium chloride also seems to be important in many environments.

(b) Changes in temperature produce greater amounts of expansion in many common salts than in many rocks. The expansion of salts can therefore damage the walls of rock capillaries, causing the rocks to split or disintegrate into grain-sized particles. The process is probably most effective where there are large **diurnal** ranges of temperature.

(c) Salt crystals growing from a solution can also exert considerable pressures on rock capillaries. The pressures exerted by sodium chloride (halite) are particularly high, although most experiments have shown that sulphates are generally more effective than carbonates, which in turn are more effective than chlorides. Solutions must be supersaturated for crystallization to occur. This can be brought about by evaporation, or a drop in temperature. Salt crystallization and other salt weathering processes are important in coastal regions, particularly where salts are carried landwards by sea breezes. They also appear to be quite active in dry polar deserts, however, despite low temperatures and evaporation. Weathering features on Ellesmere Island, for example, have been partly attributed to the effects of salt crystallization (Watts 1983, 1985). They include accumulations of loose granitic flakes around rock outcrops, tors consisting of rounded blocks of granite, and circular weathering pits up to about 0.9 m in diameter and 0.25 m in depth. These pits probably developed where slight depressions collected precipitation on the surface of the rock. Similar features have been reported on Baffin and Somerset Islands.

Heating and Cooling

The expansion and contraction of rocks as they are alternately heated and cooled may eventually cause them to break apart, particularly if they are composed of minerals that expand at different rates. This process was once thought to be very important in arid areas, though the idea has not been supported by experimental work. The results are not conclusive, however, and there is some evidence to suggest that the process may be significant in hot deserts. Fire may also cause thermal expansion and contraction.

Wetting and Drying

Rocks can also be damaged by changes in volume as a result of the absorption of water, although little is known of the processes involved. The phenomenon could be the result of ordered water adsorption, whereby water molecules become attached to the surfaces of the small capillaries within a rock. This involves the attraction of the positively charged ends of the water molecules (two hydrogen ions) to clay particles with a net negative charge. Alternate wetting and drying can stack the water molecules on top of each other like tiny bar magnets, forming a rigid, quasi-crystalline layer (Fig. 2.2). Although this water is virtually unfreezable down to temperatures of -40°C, the growth of these rigid layers can generate damaging pressures against the confining walls of small rock capillaries (Hudec 1973). It has been found that rocks contain the maximum amount of adsorbed water at temperatures between 20 and 25°C. The expansion of rocks as they adsorb water as temperatures rise, and desorb and contract as temperatures fall, may therefore eventually cause them to fail. The process is probably most effective in fine-grained, clay-rich **argillaceous** rocks, including shales, siltstones, and some dolomites.

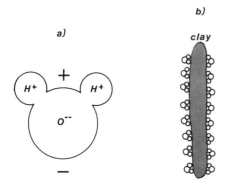

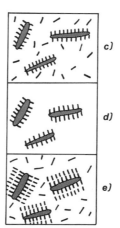

a) water molecule
b) positive hydrogen end of water molecules
   attached to negatively charged clay particle
c) water ordered around clay but disordered in
   bulk water
d) ordered water on clay after drying
e) further layers of ordered water added with
   renewed wetting

Figure 2.2
The adsorption of water molecules by clay particles (Ollier 1969).

Destructive pressures can be generated in rocks during freezing only if they have very high levels of water saturation; otherwise the approximately 9 per cent expansion can be easily accommodated within the capillaries. With the possible exception of the **intertidal** zones of coasts, very high levels of saturation are difficult to attain in the field. Argillaceous rocks, however, can become saturated with water adsorbed directly from the air, under conditions of high humidity. In Gaspé, Québec, shales in the coastal cliff were saturated (100 per cent) when the relative humidity was 100 per cent, but slates and greywackes were only 69 and 18 per cent saturated, respectively. When the relative humidity was 88 per cent, the shales and greywackes were only 79 and 7 per cent saturated, respectively. It is therefore quite possible that fine-grained, clay-rich rocks are broken and shattered by the ordered water adsorption mechanism, rather than by true frost action. It has been suggested that this process may account for the disintegration of siliceous shales on the unvegetated gravel bars of the Souris River in southwestern Manitoba (Mugridge and Young 1983).

Chemical Weathering

Changes occur to rock minerals that make them more stable under prevailing atmospheric, hydrospheric, and biological conditions. These changes largely involve the removal of their more soluble components and the addition of **hydroxyl** groups, and carbon dioxide and oxygen from the atmosphere. The chemical alteration of rocks forms clays and other materials composed of new minerals.

Water plays a particularly important role in chemical weathering. Chemical reactions are assisted by the separation of water molecules into H^+ and OH^- **ions**. The **pH** of a solution, which describes its acidity, is the negative log of the hydrogen ion concentration. Hydrogen ions are chemically very active, readily replacing other **cations** to form new compounds and combining with hydroxyls to form water.

Chemical Reactions

Chemical weathering is usually accomplished by a number of processes operating together. The main processes include:

(a) Hydrolysis: the reaction between rock minerals and the H^+ and OH^- ions of water. Silicates are largely decomposed by hydrolysis, which is the dominant primary weathering process of **igneous** rocks. The process involves the separation of metal cations from the silicate mineral structure and their replacement by the H^+ ion. Most of these replaced cations are then soluble in natural water. The process will continue as long as there are easily replaced cations, the solution is not saturated with the replaced ions, and there are still free H^+ ions available.

(b) Ion (cation) exchange: the substitution of ions in solution with those of a mineral. The common metal cations (Na^+, K^+, Ca^{2+}, Mg^{2+}) are readily exchangeable. The process is most effective in clay minerals because of their net negative charge, which cation exchange attempts to neutralize.

(c) Oxidation: essentially the reaction between a substance and oxygen to form oxides or, if water is also incorporated, hydroxides. Iron, for example, which is readily oxidized, undergoes transformation from the ferrous to the ferric, oxidized state. Oxidation probably always occurs through the presence of dissolved oxygen in water, usually in the zone above the level of permanent water saturation. It occurs most readily in alkaline environments, and in sandstones, limestones, and some shales, rather than in igneous or **metamorphic** rocks. Inorganic oxidation may result from the work of bacteria.

(d) Reduction: the opposite of oxidation, it usually occurs in water-logged conditions where there is little oxygen, although it can operate above the **water table** if there is an abundance of organic matter. Much reduction is carried out by bacteria.

(e) Carbonation: the reaction between minerals and carbonate or bicarbonate ions. When carbon dioxide dissolves in water it forms carbonic acid. Despite its weak acidity, carbonic acid plays an important role in chemical weathering, including ion exchange and the solution of carbonates (Chapter 10).

(f) Hydration: the absorption of water by a mineral so that it is loosely held by its structure. The process is important in the formation of clays. Hydration prepares minerals for alteration by other chemical processes, but it also causes considerable increases in volume, which may facilitate breakdown by physical weathering.

(g) Solution: true solution is the complete dissociation of a mineral. Solution can occur in running water or in a thin film of water. Most common elements and minerals are soluble to some extent, but gypsum, carbonates, and salt are particularly soluble (Chapter 10).

Chemical Rock Weathering

Fresh, unweathered rocks consist of combinations of minerals that result from the cooling of **magma**. Oxygen contributes almost half the weight of the Earth's crustal rocks, while silicates, which constitute a little more than one-quarter, are the most common rock-forming minerals.

Silicates • The basic structural form of the silicates is the four-sided silicate tetrahedron (SiO^4), which has one silica ion centrally located between four surrounding oxygen ions. The units are held together by the attraction between the positively charged silica ion and the negatively charged oxygen (O^{2-}) ions. Silica tetrahedra occur singly or in chains and sheets (Fig. 2.3). A mineral's resistance to weathering is largely determined by the strength of the bonding force holding the tetrahedra together. The crystalline structure is very strong, but it is weakened by the substitution of aluminum (Al^{3+}), magnesium (Mg^{2+}), iron (Fe^{2+} or Fe^{3+}), sodium (Na^+), potassium

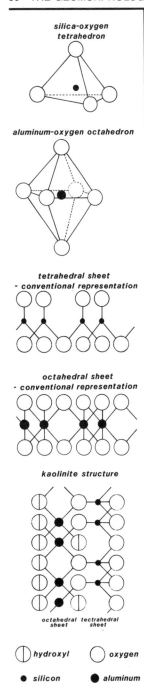

silica-oxygen
tetrahedron

aluminum-oxygen octahedron

tetrahedral sheet
- conventional representation

octahedral sheet
- conventional representation

kaolinite structure

octahedral   tectrahedral
sheet   sheet

○) hydroxyl   ○ oxygen

● silicon   ● aluminum

Figure 2.3
*Tetrahedron, octahedron,
and the structure of the
clay mineral kaolinite.*

(K^+), and other metal cations that have smaller electrical charges than silica. Hydrolysis is the main weathering process on silica minerals, although ion exchange and oxidation of the iron minerals are also important.

In general, the relative susceptibility of the rock-forming silicates to weathering is directly related to the order in which they crystallize out from cooling magma. The most susceptible minerals form at high temperatures, where Al^3 and other ions are most easily substituted into the silica tetrahedra. Olivine, pyroxene (augite), and other minerals that formed at high temperatures are therefore the least stable, and they weather most rapidly. On the other hand, quartz, as the last to crystallize, is formed at the lowest temperatures, and therefore should be most stable at the Earth's surface today.

Clay minerals • Most of the Earth's surface consists of the **unconsolidated** remains of greatly weathered rock, rather than solid, fresh rock. This material is a mixture of minerals that resisted alteration, new minerals created by the weathering processes, and organic material. Minerals created by weathering under present environmental conditions are more resistant than their ancestors, which were created from a cooling melt. The most common minerals created by weathering are the clay minerals and **amorphous** hydrous oxides of iron, aluminum, silica, and titanium. The weathering of all the common rock-forming minerals other than quartz forms clay. Clay minerals are common in soils and unconsolidated deposits, including glacial **tills**, and they have also been compacted and hardened into **sedimentary** rocks, which cover about three-quarters of the Earth's surface. Because of their small size (usually less than 0.002 mm) and negative electrical charges, clay minerals play a very active role in weathering, readily exchanging cations with metallic ions and organic **colloids**.

Clay minerals consist of sandwich-like layers of silica tetrahedra and aluminum, iron, or magnesium octahedra. These octahedra consist of a metal ion bonded to six equally spaced oxygen or hydroxyl ions. The attachment of tetrahedral silica sheets to octahedral hydroxyl sheets produces a layered structure (Fig. 2.3). Different clay minerals are formed according to the way in which these layers are combined.

Rocks • Although the composition of a mineral is important in determining how fast it weathers, other factors are also significant. These include the size, shape and degree of perfection of the crystals, the access afforded to the weathering agent, and the rate at which the weathered products are removed. The number of relevant factors is even greater for rocks, which vary according to their texture, degree of fracture, strength of bonding, **porosity**, permeability and their ability to hold water. Rock **dip**, **bedding planes**, joints, and **faults** affect the ease with which water is able to travel through a rock mass, and its ability to evacuate the soluble products. The presence of small cracks in rocks greatly increases the efficiency of chemical weathering. The position of the site is also of enormous importance,

particularly because of its effect on drainage. Attempts to rank the susceptibility of rocks to weathering are therefore constrained by the influence of
local factors, and are useful only at the most general level. In two areas of
the Canadian Shield near Montreal and Québec, for instance, it has been
found that the calc-silicate rocks and marbles were the most weathered,
while the pegmatites were the least weathered. Granite gneiss, granite, and
gneissic gabbro were moderately weathered (Bouchard and Godard 1984).
Estimates of weathering rates vary enormously. In the Sherbrooke area of
Québec, it has been calculated that postglacial weathering has lowered rock
surfaces at average rates of between 0.5 and 1.2 mm per thousand years,
with the higher rates occurring where water accumulates on low slopes and
where there is a vegetational mat (Clement et al. 1976). In the Sudbury
area, however, it has been estimated that devegetated quartzite and gabbro
bedrock exposures are currently being weathered at rates equivalent to
about 5 to 17 cm per thousand years (Pearce 1976).

Biological Weathering

Biological weathering is the result of biophysical or biochemical processes
associated with living matter. Although living organisms are probably the
most important factors in the weathering of rocks and minerals, the actual
processes are often very complex and, at best, poorly understood.

Worms and burrowing animals mix large amounts of soil, exposing new
material to weathering agencies, carrying organic material deeper into the
soil, and increasing access to air and water. It has been estimated that the
cast-making worms in a hectare of land annually pass an average of about
25,106.35 kg of soil through their bodies. Chemical solution is also
enhanced by the respiration of living creatures, which increases the carbon
dioxide content of the soil.

Most types of bacteria break down the complex carbon compounds of
vegetational debris, but some also oxidize sulphur, iron, and other minerals
for their metabolism. They are very active in reducing environments, man-
ufacturing the sulphides that are characteristic of these conditions. The
floors of heated buildings have been heaved upwards 6 to 20 cm in Sainte-
Foy, Québec and in Ottawa (Quigley et al. 1973). This has been attributed
to the oxidation of pyrite, and the precipitation in rock fractures of secondary sulphates with a greater volume than the primary minerals. Purely
chemical oxidation would be difficult in the very acidic **groundwater** in
these two areas, and it is therefore most likely that bacteria are involved in
the reactions. The presence of economically recoverable ferric iron and
uranium in solution in acidic mine water in the Elliot Lake area of Ontario
has been also been attributed to bacterial oxidization and leaching (Harrison
et al. 1966).

Algae, lichen, and fungi colonize rock surfaces, holding thin films of
water in contact with the rock and affecting the alteration of minerals by
physical and chemical processes. The most important function of plant and

animal micro-organisms, however, is to break down organic material in the soil, promoting mineral ion exchange and releasing nitrogen to plants.

Larger forms of vegetation also exert direct and indirect influences on the nature and efficacy of the weathering processes. The roots of larger plants exploit and widen cracks in rocks, hastening their disintegration. The type of vegetational cover in an area affects the removal of weathered products, and therefore the rate of weathering. Decaying vegetation promotes weathering by conserving moisture and is rich in chelating agents, which are responsible for many complex chemical effects, including the formation of organic-mineral complexes. Chelating agents are able to mobilize or remove metallic ions that would otherwise be extremely immobile, including iron and aluminum under pH conditions that would be unsuitable for the solution of these elements. Mildly acidic solutions from a present or former soil cover, for example, are thought to be primarily responsible for the selective weathering of glacial and **glaciofluvial** deposits in southern British Columbia (Bastin and Mathews 1979).

Climatic Effects

Temperature and the availability of water are the main climatic factors determining the efficacy of weathering, although other factors, including aspect, wind, cloud cover, humidity, and duration of snow cover, are also important.

Water is required for frost action and chemical reactions, as well as to evacuate the weathered products. Reactions can continue if the products of weathering are removed, but if they remain the reactions will terminate and an equilibrium state will be attained. For instance, high runoff may partly explain the fairly rapid removal of chemically weathered silica from three small watersheds in the Cordillera (Slaymaker and McPherson 1977). But the relationship between precipitation and weathering efficacy is complex. Seasonal changes in precipitation may be necessary for some processes, whereas a more equitable distribution may be needed for others. The amount of evaporation, relative to precipitation, influences the movement of solutions through the soil, the type of soil that develops, the type of clay minerals present, and the role of salt weathering.

It is generally assumed, although unproven, that fairly rapid fluctuations in temperature about the freezing point are more suitable for frost action than are fluctuations of greater amplitude but lower frequency. Chemical reactions tend to be faster at higher temperatures, and biological activity is also greater. On the other hand, the increase in the chemical reaction rate with higher temperatures may be offset by a corresponding reduction in the solubility of oxygen and carbon dioxide in water. Former climatic conditions also have to be considered. Deep chemical weathering of gabbro on Mount Megantic, and elsewhere in Québec, may have been the result of the warmer climate of the last interglacial stage, and the opening of joints and fractures by deep frost action during the **Wisconsin glacial** stage

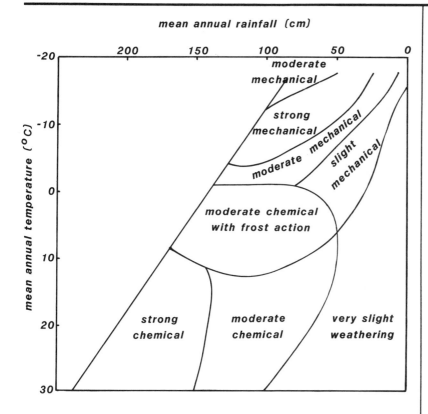

mean annual rainfall (cm)

Figure 2.4
The relationship between climate and type of weathering according to Peltier (1950).

(Clément and De Kimpe 1977). Ancient weathered rock or **saprolite** has also survived glaciation and been incorporated into the soil of southern Québec and parts of New Brunswick and Nova Scotia (McKeague et al. 1983, Bouchard 1985).

There have been several attempts to model the distribution and efficiency of weathering processes based upon simple climatic criteria. Because of the complex and poorly understood nature of the relationships between climate and weathering, however, these models can provide only a crude representation of weathering patterns on the global scale (Fig. 2.4). Effective frost action, for example, requires sufficient moisture to attain critical levels of saturation, and, possibly, temperatures that fluctuate rapidly about the freezing point. The frequency of freeze-thaw cycles is greater in southern than in northern Canada (Fig. 2.5) (Fraser 1959), while the length of the freezing period and the degree of coldness attained are greatest in the north. In tundra regions, the percolation of water is inhibited by frozen ground, and there appears to be less pronounced chemical weathering in the drier and colder regions of Canada than in hotter and wetter areas. The role of chemical weathering has been documented in the eastern and central Arctic, although the rates are very low and the effects are probably secondary to those of physical weathering. Dolomitic surfaces on the eastern side of Hudson Bay, for instance, have been lowered by chemical weathering at average rates of about 6 mm per thousand years, since the area emerged from beneath the Tyrrell Sea (Chapter 5) (Dionne and Michaud 1986). Iron oxide stains on rocks have been noted in many parts of northern Canada, and a protective surface coating has formed on joint blocks through the solution, migration, and **deposition** of silica. This coating is generally deteriorating, however, and may therefore have originally developed when the climate was warmer than today.

Rapid weathering has had an important role in the development of **badlands** in several parts of southern Alberta (Campbell 1987). Badlands form in fairly weak rocks, particularly where there is a period of brief, concentrated runoff from heavy summer rainfall, and a long period of frost in winter. Under these conditions, the rock is weathered and eroded so rapidly that soils and vegetation cannot develop. Runoff is therefore able to create a deeply incised surface with a dense maze of rills and channels. Present rates of **denudation** in Dinosaur Provincial Park on the banks of the Red Deer River, for example, are about 4 mm yr^{-1}.

The weathering zone concept has been applied to the study of the Pleistocene glacial history of eastern Canada, from Newfoundland to eastern Baffin Island (Clark 1988). Weathering zones are areas of bedrock, till, or other surficial deposits that can be distinguished according to differences in their degree of weathering. It is believed that these variations reflect the age of the deposits, although other interpretations are possible; of the four zones that have been recognized along the northern and central Labrador coast, the lowest three have been interpreted as representing separate glaciations.

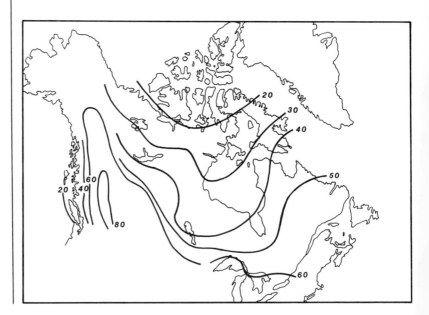

Figure 2.5
Average number of freeze-thaw cycles per year in Canada (Fraser 1959).

Mass Movement

3

Mass movement (**mass wasting**) is the downslope movement of rock and soil in response to gravity. The term does not include mass transportation by other agents such as snow, ice, water, and air, although the distinction becomes rather arbitrary in some cases. Gravity exerts **stress** on all slope materials. This increases with the gradient and height of the slope, and with the unit weight of the material. Additional stresses are generated in the surface layers by alternate freezing and thawing, shrinking and swelling, and thermal expansion and contraction. Downslope movement begins when the stress becomes greater than the resisting strength of the material. The stress and strength are usually about equal, and movement is generally quite slow. Sudden movements can occur, however, when there is a rapid increase in stress, or a rapid decrease in the strength of the material. The former could be brought about by the additional weight of a new building, or by vibrations from earthquakes, traffic, machinery, blasting, or thunder. Weathering of the surface layer causes a gradual decrease in strength, but sudden decreases can be produced by an increase in the amount of water in the material during periods of heavy rainfall or ice- and snow-melt. The more important effects of water include the undermining of slopes by the washing away of beds of soluble or loose material; the softening of **colluvium**, helping it to flow; the swelling of clay-rich rocks (Chapter 2); and the generation of high pressures in the pores and clefts of rocks and soil. One of the most common reasons for slope movements is the removal of material from the base, or foot, whether by glacial, fluvial, or coastal **erosion**, by a previous slope failure, or by the work of man.

Mass movements have been classified in several ways, using a variety of criteria. In the classification used in this discussion, five main classes are defined on the basis of the type of movement, while a sixth represents movements that are a combination of two or more types (Table 3.1). Further division is based upon the type of material before movement, or before the type of movement changes. Bedrock is distinguished from loose or **unconsolidated** material, which is further divided on the basis of the size of the material.

Slope movements in Canada are common in the Cordillera, on the sides of river valleys on the western Interior Plains, in the St Lawrence and James Bay Lowlands, along the shores of the Great Lakes, and in the permafrost regions of the north (Mollard 1977, Eisbacher 1979a, Eisbacher and Clague 1984, Johnson 1984a, Cruden 1985, Heginbottom et al. in press, Cruden et al. 1989). Submarine slope movements, including massive landslides triggered by earthquakes, also occur offshore on the continental shelves. The type of slope movement that takes place in a particular area depends upon geologic, topographic, and climatic factors, and the effects of human activity such as mining, logging, and urban development. Many types of mass movement threaten life and property, and may block economically vital transportation routes for considerable periods of time. It has been estimated that annual losses in Canada from landslides, including flows, may be of the order of one billion dollars.

Table 3.1   Classification of slope movements (Varnes 1978).

TYPE OF MOVEMENT			TYPE OF MATERIAL		
				UNCONSOLIDATED	
			BEDROCK	COARSE	FINE
FALLS			Rock fall	Debris fall	Earth fall
TOPPLES			Rock topple	Debris topple	Earth topple
SLIDES	Rotational	Few Units	Rock slump	Debris slump	Earth slump
	Translational	Many Units	Rock block slide	Debris block slide	Earth block slide
			Rock slide	Debris slide	Earth slide
LATERAL SPREADS			Rock spread	Debris spread	Earth spread
FLOWS			Rock flow (deep creep)	Debris flow[1] (soil creep)	Earth flow[2] (soil creep)
COMPLEX			Combination of two or more types of movement		

[1]Includes debris avalanches, solifluction, and soil creep
[2]Includes rapid earth flows in quick clay, mudflows, and wet and dry sand and loess flow.

Falls and Topples

Falls are rapidly falling, leaping, bouncing, or rolling descents of material, mainly through the air, from cliffs and other steep slopes (Fig. 3.1 a). Blocks or fragments of rock, debris, or earth become detached from the outer part of a cliff, often because of water freezing in cracks and other openings and enlarging them (Chapters 2 and 7). Falls also occur where erosion at the base of a slope forms an overhanging ledge. This is common where coastal cliffs are exposed to vigorous waves, although similar results can be achieved by **stream** or glacial action.

Slabs of rock topple or overturn as a result of forward tilting or rotation (Fig. 3.1 b,c). This is particularly common where distinct columns of rock are defined by **joints**, **cleavage**, or **bedding planes**. Rocks topple where the structure is unsuited for sliding, and, given the same rock type, they tend to occur on slopes that are steeper than those characterized by sliding.

Small rockfalls in the Rockies are common where glacially oversteepened slopes are weakened by frost action, particularly on shaded, leeward slopes facing between north and northeast (Gardner 1983). Rockfalls are most frequent between November and March in the Fraser Canyon, when fluctuations in temperature and saturation of the rocks provide the most suitable conditions for effective frost action. Most rockfalls occur in the spring in the uplands of Baffin Island and in Jasper National Park, although there is also a second, much lower maximum in the fall in the latter area (Luckman 1976, Church et al. 1979). On the other hand, no clear seasonal pattern has been found in the Foothills or in the Lake Louise area of the Rockies. Although most rockfalls are in rugged, remote areas where they have only occasionally threatened human life, eighty-five people died and several houses were destroyed by rockfalls in Québec City between 1836 and 1889.

Thermo-erosional falls occur along river banks and coastal bluffs consisting of ice-rich **sediment**, or ground ice. As the water melts the ice, the formation of a deep notch at the water level eventually causes the overlying material to collapse. Thermo-erosional falls are common in the ice-rich sediments along the coast of the western Arctic, and along the banks of streams and the channels of deltas.

Slides

Slides are slope movements involving material slipping over one or more surfaces. They are particularly common in wet, **tectonically** active and earthquake-prone areas with strong **relief**. Glaciation and rapidly downcutting streams provide suitably steep slopes, often exposing geological discontinuities – including bedding, cleavage, or joint planes – **dipping** steeply outwards. The distinction is usually made between translational slides along essentially straight or planar surfaces, and rotational slumps along curved surfaces.

Figure 3.1.
*Falls, topples, and slides
(modified from Varnes
1978).*

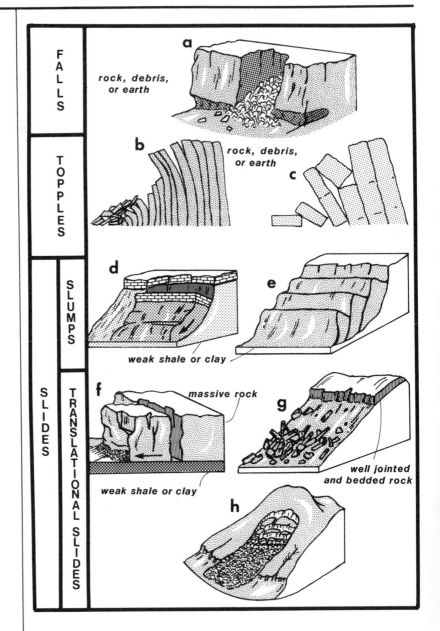

Translational Slides

The planar **rupture surface** in a typical translational slide is roughly parallel
to the ground surface, and at fairly shallow depths. In rock, the rupture
surface often lies along discontinuities, especially where they are occupied
by a clayey filling. In **cohesive** soils, the slip surface often coincides with
a particularly weak layer, or with a change from soft to hard material.

There are several types of translational slide. In block slides (or block glides), the material moves as one or a number of closely related units with little fragmentation (Fig. 3.1 f). They are common in the Salmon Valley of central British Columbia, and have also been reported from Fundy National Park in New Brunswick, where marine undercutting has caused large blocks of sandstone to move seawards over the impermeable rocks below. In rock or debris slides (Fig. 3.1 g and h, respectively), the material disintegrates into many units and as the velocity and water content increase, they may become debris avalanches with some of the characteristics of flows.

Large slides are common along the sides of valleys in the Interior Plains and on the slopes of uplands such as the Horn Plateau, the Birch and Caribou Mountains, and the Swan Hills. Many of the valleys were cut by meltwater flowing at the margins of the ice sheets or from ice-dammed lakes during the latter stages of glaciation. Rapid meltwater erosion of deep, steep-sided valleys caused extensive landsliding at that time. Although there has been little further downcutting in the last 6,000 years, valley widening and river meandering have eroded the **toe** of the slopes and caused new landslides. Many landslides are inactive today, but they can easily be reactivated, posing threats to bridges, roads, and dams. Movement, however, is generally quite slow. In Saskatchewan, for example, the large Denholm slide near North Battleford moved only 35 mm per year on average over the last 11,000 years, and it is now probably dormant, while the **scarp** of the much smaller Beaver Creek slide is presently **retrogressing** by about 1 m per year.

Slides in the Interior Plains are particularly common in bentonitic marine clay shale, silty shale, and claystone. They can extend for several kilometres along a valley side, and be more than a kilometre in length from scarp to toe with gradients of between 4 and 9.5°. In most areas they consist of roughly wedge-shaped units defined by planar rupture surfaces. The lowest, or **basal**, surface is usually essentially horizontal and coincident with a particularly weak bed, such as a thin seam of bentonite. A few slides, however, are more characteristic of slumps than translational slides; the Beaver Creek slide south of Saskatoon consists of up to four rotating and subsiding blocks.

Slumps

Slumps take place along curved surfaces that are concave upwards, and usually at greater depths than translational slides. Thick, homogeneous deposits of clay or shale are particularly susceptible. Slumps can also take place in harder rocks, but their shape is affected by structural influences and the effect of beds of varying resistance to **shear**. Rotation lowers the **head** of a slumping mass and raises its toe. When a block subsides, the steep scarp is left unsupported and water may collect in the reversed slope at the head of the slump block. This encourages further slumping, until a more stable slope of low gradient is attained. One may therefore distinguish between

(a) simple rotational slips, in which an essentially coherent unit moves along a single surface (Fig. 3.1. d); and

(b) multiple rotational slips, in which movement of blocks takes place along several curved slip surfaces. Each of these surfaces is tangential to a single, usually deep-seated, rupture surface (Fig. 3.1 e).

Deep-seated rotational slumps are much more numerous than shallow, planar rockslides in the Skeena Mountains of north-central British Columbia. Slumps are also common in the Foothills of the Rocky Mountains, in the St Elias Mountains, and in the Interior Plateau of British Columbia and its counterparts to the north in the Yukon (Fig. 1.9).

Lateral Spreads

The main characteristic of spreads is the lateral extension or spreading of fractured material. Although bedrock extension can take place on the crest of ridges without a **plastic** zone or well-defined basal fracture surface below (Fig. 3.2 a), in some cases a layer, such as bedrock (Fig. 3.2 b, c) or clay (Fig. 3.2 d), is extended or stretched by the liquefaction or plastic flow of the underlying material. This can be the result of plastic material being squeezed out by a heavy overlying layer as it subsides. Movement is usually complex, however, involving elements of translational sliding, rotation, and flow. An extensive area of lateral spreading has been observed in the Liard River basin near Fort Nelson in northwestern British Columbia (Gerath and Hungr 1983).

Figure 3.2.
Types of spreads and flows (modified from Varnes 1978).

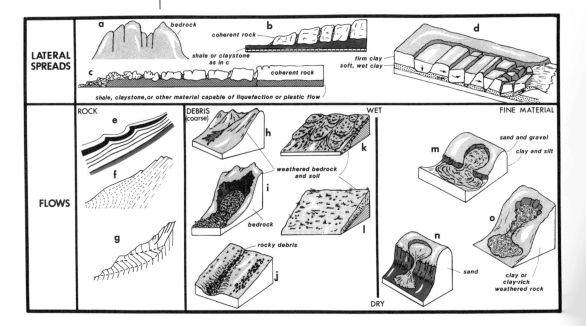

Flows

The movement in true flows is similar to that in a **viscous** fluid, with velocity declining with depth within the flowing material. Water is necessary for most types of flows, although dry flows can develop from rockfalls and rockslides. Flows usually terminate in gently sloping areas or at obstacles such as fast-flowing rivers.

Creep

Creep may be defined as the imperceptibly slow, flowing deformation of a slope. Two kinds of creep can be distinguished. Mantle creep is restricted to a surface zone where there are seasonal variations in temperature and moisture, whereas deep-seated or mass creep, which is driven only by gravity, can deform rock slopes down to considerable depths.

Mantle creep • Mantle creep is the shallow, largely seasonal movement of a weathered surface zone. It usually takes the form of soil creep, particularly at depths extending down to less than 1 m below the surface (Fig. 3.2 l). The evidence for active soil creep includes the tilting and displacement of posts and other structures, the slow movement of isolated boulders down gentle slopes, the downslope bending of weathered beds of rock, the accumulation of soil on the upslope side of obstructions, and possibly the occurrence of curved tree trunks that are concave upslope. Soil creep may result from alternate expansion and contraction of the soil caused by variations in temperature and moisture content. Measurements in humid temperate environments suggest that movement in the upper 50 mm of soil averages about 1 mm per year, although variations occur because of differences in such factors as slope angle and moisture content.

Another form of mantle creep affects accumulations of rock fragments at the foot of steep cliffs. Talus creep involves the very slow downslope movement of the surface layers, at least in part because of alternate expansion and contraction as a result of fluctuations in temperature. This may take the form of freeze-thaw action in cold periglacial regions.

Mass rock creep • Deep-seated continuous or mass rock creep is the slow but large-scale movement of fractured bedrock along poorly defined surfaces. These movements are the result of stresses generated by the weight of the material above, and they can occur down to depths of 300 m. Creep can cause rocks to bend, bulge, and fold on slopes (Fig. 3.2 e,f,g). Scarps facing upslope and grabens or troughs on top of hills in the Coast Mountains of British Columbia are indicative of gravitational deformation and partial collapse of entire mountain ridges, possibly because of the removal of lateral support from valley sides by glacial erosion (Fig. 3.3) (Bovis 1982). Continuous creep is especially pronounced where mudstone or some other soft, visco-plastic rock is overlain by a more resistant, rigid rock, but it can also occur in more homogeneous material. Typical rates of movement are several centimetres per year, although much faster rates have been recorded. Rock creep can precede rapid or catastrophic slope movements, or it can continue for many years without any acceleration.

Mass rock creep or sagging is considered to be one of the most important types of mass movement in the Cassiar – Columbia zone of the western Cordillera (Fig. 1.9). Broken rock slabs move slowly downslope along numerous internal dislocation zones, producing a distinctive scarp at the rear. The Downie slide north of Revelstoke in southern British Columbia

Figure 3.3.
Uphill-facing or antislope scarps along Central Creek in the Coast Mountains of British Columbia (Bovis 1982).

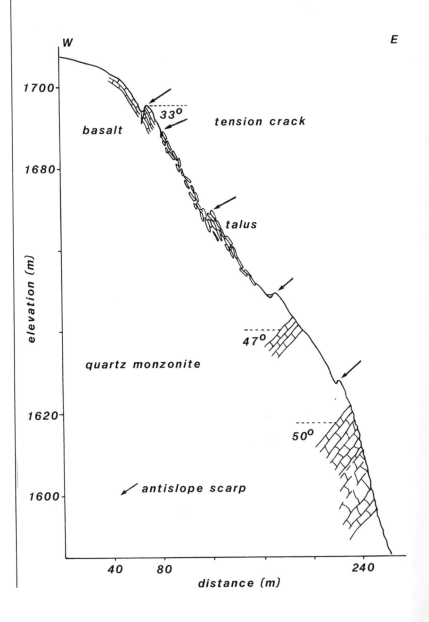

is a large example of slope sagging, with an estimated volume of between 1 and 2 billion cubic metres, and a surface area of about 9 km². Movement began at least 6,600 years ago, and it continues at rates of a few centimetres per year.

Avalanches

Avalanches are rapid flows of material downslope in mountainous areas. The distinction is made between snow avalanches, which consist largely of snow and ice, and debris avalanches, which involve large masses of rock fragments.

Snow avalanches • Snow avalanches in the Cordillera play an important, if somewhat neglected, role in moving debris downslope. Their sources are usually bare, unvegetated depressions above the tree line, but they also occur in the wooded areas at lower elevations. They are generally 'clean', but a considerable amount of debris may be included where they involve the whole depth of the snowpack or run into snow-free areas. The **geomorphic** effects of snow avalanches include the modification of talus slopes, and the deposition of boulder tongues and cone-shaped debris accumulations in their run-out zones (Luckman 1977, Butler 1989).

Debris avalanches • When very large amounts of broken rock fragments are produced by a rockslide or a fall, the debris assumes a streaming or flowing motion (Fig. 3.2 i). The exact mechanisms have not been determined, but it has been suggested that the debris flows in or on a cushion of compressed air, a dense cloud of dust, vaporized **interstitial** water, or wet mud. Avalanche debris surmounts major obstacles as it advances far beyond the foot of slopes, rising more than 100 m up the opposite sides of valleys. Movement usually begins with translational sliding, followed by falls and flows. Flow rates are commonly between 50 and 330 km hr⁻¹. North of Vancouver, for instance, maximum velocity has been estimated at between 54 and 72 km hr⁻¹ (15 and 20 m sec⁻¹) in the 1963 Dusty Creek Slide (Clague and Souther 1982), more than 72 km hr⁻¹ in the nineteenth-century Rubble Creek Slide (Moore and Mathew 1978), and up to 360 km hr⁻¹ in the 1959 Pandemonium Creek Slide (Evans et al. 1989).

Debris avalanches in western Canada are particularly common in the southern Rocky Mountains, the Coast Mountains, and the Mackenzie Mountains (Cruden 1976, Clague 1987). They tend to occur in well-jointed rock dipping at angles of between 25° and 40°. The rupture surface is usually roughly parallel to the original mountain slope, and the direction of slip movement is nearly always approximately normal to **anticlines**, **synclines**, **thrust faults**, and other regional geological structures. They seem to occur most often in Quaternary volcanic rocks, and in valleys that have been overdeepened and oversteepened by stream and glacial erosion. Movement may be triggered by heavy precipitation, frost action, or large earthquakes. For example, the high incidence of debris avalanches and other types of landslide in the Kluane Lake area of the Yukon has been partly attributed to the high seismicity of the area.

The Frank Slide, in the Crowsnest Pass of southwestern Alberta, is the most famous debris avalanche in the Canadian Rockies (Fig. 3.4). At 4:10 a.m. on 29 April 1903, about 33 million cubic metres of rock slid down from the eastern face of Turtle Mountain. In probably less than 100 seconds, debris had crossed the valley floor, killing 76 people in the coal-mining town of Frank, and come to rest on the opposite side up to 120 m above the valley floor. Other, more ancient debris avalanches were even larger: the Maligne Lake Slide near Jasper had an estimated volume of 500 million cubic metres, and a horizontal distance of travel of 5.5 km (Fig. 3.4).

Many debris avalanches in the Rockies probably occurred immediately after the ice had retreated from the valleys. In one survey, over 900 large rockslides were identified within an area of about 60,000 km^2 in the Albertan Rockies. Most of them are ancient, however, and if it can be assumed that local coal-mining was at least partly responsible for the Frank Slide,

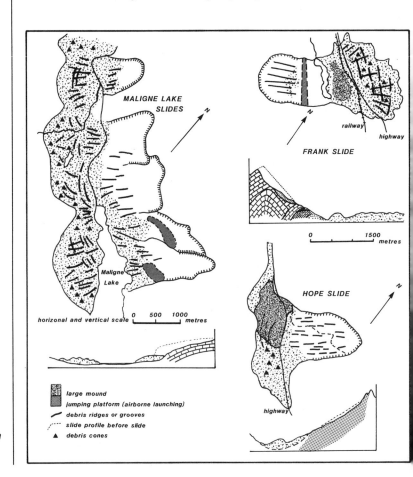

Figure 3.4.
Three debris avalanches in the western Cordillera (after Mollard 1977).

then a slide at Brazeau Lake in 1933 is the only known example of a natural slide in the Rockies in historic times.

Because of sparse vegetation and earthquake activity, large slope movements are quite numerous and conspicuous in the subarctic Mackenzie Mountains (Eisbacher 1979b). Sometime in the last 200 years, more than 500 million cubic metres of debris slid down the north side of one valley at speeds estimated to have been at least 350 to 400 km hr^{-1}. Debris dammed the valley, creating Avalanche Lake, and extended up the opposite side to about 500 m above the valley floor. Another large avalanche in this area, involving about 5 to 7 million cubic metres of debris, was triggered by the Nahanni earthquake in October 1985.

There have also been many catastrophic debris avalanches in the Coast Mountains and Cascade Range of British Columbia. The 1965 Hope Slide (Mathews and McTaggart 1969) in the Cascades had a volume of about 47 million cubic metres (Fig. 3.4). This was the largest of the eighteen debris avalanches that are known to have occurred in the Cordillera since 1855. The Hope Slide killed four people and buried a 3 km section of British Columbia Highway 3 to a maximum depth of 75 m. A debris avalanche mixed with snow killed about 60 people in 1915 in the Jane Camp (Britannia Mine) in the Coast Mountains. Four more people were killed in 1975 in the Coast Mountains in the Pylon Peak Slide, a debris and glacial ice avalanche caused by the weight of Devastation Glacier and the effect of glacial meltwater on the underlying surface (Mokievsky-Zubok 1977).

Earthflows

There are several types of mass movement involving very wet to dry, fine-grained materials. They include rapid flows of dry loess and sand (Fig. 3.2 n), and rapid, wet flows of sand or silt. Earthflows range from slow to rapid movements of plastic or fine-grained, non-plastic material.

Rapid earthflows • Because the highly sensitive quick-clays of eastern Canada have a tendency to liquefy, they are particularly susceptible to rapid earthflows (Gagnon 1972). There are extensive deposits in the Ottawa and St Lawrence River Valleys, in the Saguenay River–Lac Saint-Jean region, along the eastern coasts of James Bay and the lower part of Hudson Bay, and along the lower Hamilton River Valley west of Goose Bay, Labrador (Fig. 3.5). Sandy terraces formed where the streams cut down into these sediments.

Most earthflows in the St Lawrence Lowlands take place where late and postglacial clay and silt, deposited in the Champlain Sea between about 12,400 and 9,300 years ago, were covered by a layer of fluvial or deltaic, fine- to medium-grained sand with some gravel. The sandy layer reduces surface runoff, causing the underlying sediments to become saturated with water. This transforms the clay and silt from a brittle solid into a dense liquid, breaking up the stiff upper layer into strips that tend to subside as

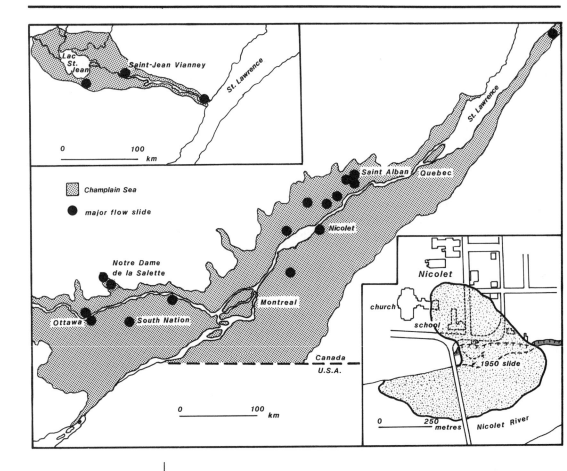

Figure 3.5.
Rapid earthflows and lateral spreads in the sensitive marine clays of southeastern Canada (after Parkes and Day 1975, and Béland 1956).

they are carried along. Earthflows can also be triggered by stream erosion at the base of the terraces, or by vibrations from earthquakes or heavy traffic. The surface skin remains in a fairly solid state in many of these movements, and they can be classified as lateral spreads (Fig. 3.2 d). In other cases, however, the entire mass is liquefied, as in an earthflow. Movements of several hundred metres can occur within a few minutes. Earthflow craters are semicircular in plan, with steep head scarps and flat or gently sloping floors (Fig. 3.2 m).

More than 750 earthflows larger than 4,000 m² in area have been identified in the lowlands of Québec and Ontario, along with thousands of smaller examples (Parkes and Day 1975). A slide at Notre Dame de la Salette in Papineau County, Québec, killed 33 people in 1908, and another at St Alban in Portneuf County in 1894 involved the loss of 6.5 km² of land. There have also been several catastrophic earthflows in the post-war period. The 1955 earthflow at Nicolet, a small town about 129 km southwest of Québec, killed three people and destroyed several buildings, including part of a large church (Fig. 3.5) (Béland 1956). The crater was 213 m long and 122 m wide, and it occupied an area of 22,000 m². There were two major earthflows in May 1971, during periods of heavy rainfall and towards the end of the snow-melting season. The one at Saint-Jean-Vianney, about 10 km west of Chicoutimi, Québec, killed 31 people and destroyed 43 houses in a new residential development. This slide began in the **zone of depletion** of a much larger slide that occurred several hundred years ago. Liquefied debris flowed at about 26 km hr⁻¹ along the valley, carrying 34 houses, a bus, and an unknown number of cars, and destroying a bridge that lay in its path. The second major incident, the South Nation River earthflow, occurred about 48 km east of Ottawa, several kilometres north of Casselman, Ontario. The roughly semicircular depletion zone along the stream bank was about 640 m wide and extended about 490 m back from the river. Narrow grass-covered blocks, up to 200 m in width, remained intact as they subsided 8 to 10 m in the upper part of the flow, but there was little backward or forward tilting, and therefore little rotational movement.

Rapid earthflows also occur in the Interior Plateau of British Columbia, where major rivers have cut into **glaciolacustrine** silt. Flows have been generated by heavy rainfall, or in some cases by water from irrigation. In 1905, for instance, 15 people were killed near Spences Bridge in the Thompson Valley by a rapid earthflow that has been attributed to the irrigation of a valley bench.

Slow earthflows • Somewhat drier and slower earthflows, which often occur as the spreading or bulbous toes of slumps, are common where there are moderate gradients and clay or weathered clay-bearing rocks (Fig. 3.2 o). Such movements are often protracted over many years. Large, slow-moving earthflows are common in the weathered and altered sedimentary and volcanic rocks in the deep valleys of the western part of the Interior Plateau of British Columbia (Fig. 3.6) (Bovis 1985). The earthflows in this area are up to 6 km in length, and they can involve several million cubic metres of material. Movement has taken place over several thousand years in many cases, at rates varying from a few centimetres to half a metre per year. One of the largest examples is the Drynoch Earthflow, which resembles a valley glacier flowing into the Thompson River, about 180 km northeast of Vancouver. The flow has a volume of about 17 million cubic metres and is over 5 km in length, although it attains a maximum width of only 670 m. Movement began between 3,000 and 6,000 years ago, and in modern times it has been as high as 3 m per year in some parts. The earthflow is crossed by the Trans-Canada Highway and the main CPR line to Vancouver. Disruption of these transportation routes has been occurring for more than one hundred years, but it is now contained by the drainage of water from the flow.

Debris Flows

Some classifications make a distinction between debris flows and mudflows

Figure 3.6.
The Pavilion earthflow in the Interior Plateau of southwestern British Columbia (Bovis 1985).

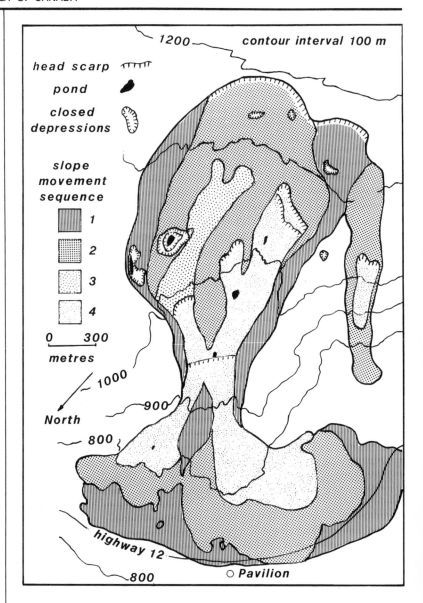

head scarp

pond

closed depressions

slope movement sequence

1

2

3

4

0   300
metres

1000

900

800

North

800

1200   contour interval 100 m

highway 12

800   ○ *Pavilion*

on the basis of particle size, mudflows containing a higher proportion of fine-grained sand, silt, and clay. Debris flows and mudflows travel considerable distances in confined channels, depositing their sediment in large cones or fans (Chapter 8) in areas of lower gradient, often at the foot of a mountain front (Fig. 3.2 h).

Debris flows and related events have killed at least 160 people in western Canada; they have washed out and blocked bridges, roads, and railways, and flooded and destroyed houses and other buildings (Evans 1982, Eisbacher and Clague 1984, VanDine 1985). Most of the fatalities were probably the result of flows triggered by human activity. Until now, development has been largely restricted to areas considered to be safe from slope movements, but the scarcity of flat land available for urbanization means that building is increasingly taking place on alluvial fans and potentially unstable slopes.

Debris flows are usually the result of heavy seasonal rainfall or snowmelt, which mobilize **glaciofluvial** deposits and colluvium in valleys and on mountain slopes. Although they sometimes develop directly from bedrock movements when water is added to the moving mass, in most cases they occur when bedrock debris blocking or constricting a stream channel is mobilized during periods of heavy rainfall or overflow. Many other factors can help to trigger slope movements. Vegetation helps to hold sediments together and controls the runoff of water. Debris avalanches and debris flows may therefore be the result of the clearing of vegetation for urbanization or for logging; logging debris may also temporarily block stream flow. Flows may also result from the saturation of the soil by irrigation and the watering of lawns, discharge from septic tanks, and leaking water mains, swimming pools, and storm drains. Glacially induced debris flows in mountainous regions can also be generated by the rapid draining of

(a) ice-dammed and ice-marginal lakes, when lake water pressure exceeds the strength of the ice dam;

(b) lakes dammed by glacial moraines; and

(c) water pockets in and under the ice.

These outburst floods often occur where an advancing glacier blocks a tributary valley. Most outbursts are largely water, but debris flows develop if there is a steep, narrow, and sediment-filled valley below the lake. In the early 1970s, for example, the draining of a moraine-dammed lake in the Klattasine Creek Basin in the southern Coast Mountains of British Columbia produced a rapid debris flow, which travelled for about 8 km and deposited sediment with large boulders to a depth of up to 20 m.

The distinction can be made between open-slope debris flows on steep slopes and debris flows that become channelized, although the former may quickly cut a V-shaped gulley. Open-slope debris flows are common on the steep walls of fiords in the Coast Mountains and on the Queen Charlotte Islands. A debris avalanche and open-slope debris flow destroyed three houses and killed eight people in 1957, on the outskirts of Prince Rupert in northwestern British Columbia. The term 'debris torrent' is often used to describe channelized debris flows in the western Cordillera. These are rapid flows of saturated, largely coarse-grained organic and inorganic material, confined by pre-existing channels. They occur in areas of high relief and high precipitation, especially in the coastal mountains of southern British Columbia, where they pose a considerable danger to urban development. Debris torrents are major hazards around mountain creeks and on the alluvial fans where the debris is deposited. In July 1988, heavy rainfall resulted in numerous debris flows that closed the Alaska Highway in northern British Columbia and the southern Yukon (Evans and Clague 1989). In July 1983 the Trans-Canada Highway and the Canadian National and Canadian Pacific Railways were severed by debris flows in the Hope-Chilliwack and Revelstoke-Rogers Pass areas of southwestern British Columbia. A mountainous area north of Vancouver containing 26 stream basins has experienced at least 14 debris torrents in 25 years. In 1964, in the logging settlement of Ramsey Arm, about 200 km northwest of Vancouver, three people were killed by a debris flow initiated by a debris avalanche with its source on logged terrain several kilometres inland.

The continuing development of debris fans, which provide gentle slopes in otherwise rugged terrain, suggests that destruction of property and loss of life will increase in the future. One of the worst disasters occurred in 1921, on a fan-delta about 40 km north of Vancouver. Collapse of a stream bank during heavy rainfall temporarily dammed Britannia Creek, and when the landslide dam eventually broke, a deluge of water, debris, and logs roared into the town of Britannia Creek, shearing buildings from their foundations and killing 37 people. There was also considerable damage, though no loss of life, in 1973 and 1975 at Rumble Beach, Port Alice, on northern Vancouver Island, when channelized debris flows extended onto the fan on which the town is built. A mixed slush avalanche and channelized debris flow, triggered by heavy rainfall onto an exceptionally thick cover of snow, killed seven people in 1965 in the town of Ocean Falls, 470 km northwest of Vancouver. Debris flows also occur within the urbanized Vancouver region during periods of heavy fall and winter rainfall, particularly where debris avalanches, carrying Pleistocene sediments, soil, and uprooted vegetation, temporarily block gullies that are swollen with water.

Periglacial flows

Several aspects of cold periglacial regions (Chapter 7) are particularly conducive to flowing movements of weathered debris, and it may be that mass movement is most intense and efficient in these environments (McRoberts and Morgenstern 1974, Lewkowicz 1987). Such aspects include the presence of frozen ground beneath the surface, which inhibits the downward percolation of meltwater supplied by snow, ice, and ground ice, as well as providing a natural, water-lubricated slip plane; the alternate freezing and thawing of the wet debris; and vegetation that tends to be too sparse to prevent its movement down even very gentle slopes.

Gelifluction and related slow flows • The term 'solifluction' literally means 'soil flow', and it has been used to refer to a very wide range of mass movements in a variety of climatic environments. The term 'gelifluction' is more specific in that it implies the existence of a periglacial environment. The term is therefore increasingly used in preference to 'solifluction' to refer to the movement of thawed surface material over permanently, seasonally, or even **diurnally** frozen ground (Fig. 3.2 k).

Gelifluction occurs when a shallow surface layer is soaked by meltwater that is unable to penetrate the frozen material below. The water increases the weight and reduces the internal friction and cohesion of the thawed layer, thereby reducing its shear strength. Rates of movement are probably between about 0.5 and 4.0 cm yr^{-1} in most areas. Movement appears to be greatest on slopes of between 5 and 20°, although it can take place on slopes as low as one degree. Movement is normally restricted to depths of no more than about 0.5 m.

'Frost creep' refers to the movement of sediment down a slope as a result of alternate freezing and thawing. When the ground freezes, growing ice crystals heave it upwards, perpendicularly to the freezing surface. Upon thawing, the ground on a sloping surface settles slightly downslope of its previous position, although it rarely settles exactly vertically. Frost creep therefore differs from normal mantle or soil creep only in that, in this case, the cycle of expansion and contraction is generated by alternate freezing and thawing. Rates of movement are probably of the order of centimetres per year. Seasonal heaving in the Foothills of the Rocky Mountains, for example, displaces the surface by 23 to 45 mm yr^{-1}, accounting for between 36 and 63 per cent of the total movement of the surface in this area (Smith 1987a). Frost creep and gelifluction frequently operate together and may be difficult to distinguish. According to local circumstances, in one year creep could dominate on one part of a slope and gelifluction on another, while their relative importance is reversed in the following year (Fig. 3.7).

Needle-ice (pipkrake) usually consists of clusters of needle-like ice crystals, up to several centimetres in length, growing perpendicularly upwards from the ground surface. At night, clusters of growing ice needles lift pebbles and finer material up from freezing surfaces. The thawing or breaking of these needles then causes the particles to settle or roll to lower elevations. Needle ice therefore produces a movement of surface particles similar to that produced by seasonal frost creep, although it is much faster. The nocturnal growth of needle ice in the Foothills of the Rocky Mountains

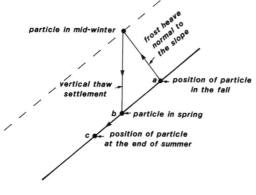

ab downslope movement due to frost creep
bc downslope movement due to gelifluction

Figure 3.7.
Movement of a particle on a slope by gelifluction and frost creep (after Lewkowicz 1989).

is responsible for an average frost heave of about 10 mm. It occurs only about 20 times a year, however, and its geomorphic significance therefore appears to be limited (Smith 1987a).

Skin flows • Skin flows (active layer or detachment failures) involve the rapid movement of an unfrozen layer of soil and vegetation over ice-rich permafrost. Individual movements are shallow and often long and narrow, although the flows can coalesce to form broad sheets. They develop on steep or gentle slopes that are lightly wooded or burnt-over, particularly when they face to the south or west. Many skin flows in the Mackenzie Valley are triggered by periods of heavy rainfall. Skin flows are probably more common in the western Arctic than in the eastern or central Arctic, where much of the surface consists of hard bedrock, or sands and gravels containing little ground ice. Similar movements known as active layer glides involve the sliding, as opposed to the flow, of the thawed active layer over a sloping permafrost surface.

Bimodal flows • The thawing of ice-rich sediments or massive ground ice can produce bimodal flows (thaw slumps, ground-ice slumps, or thaw-flow slides). Bimodal flows have steep (30 to 50°), semicircular or bowl-shaped scarps at the rear, and gently sloping (3 to 14°) terminal lobes or tongues (Fig. 3.8). They therefore resemble the earthflows of more temperate regions, but because they result from the melting of large amounts of ground ice, the scars often appear disproportionately large when compared with the amount of debris they contain.

Bimodal flows develop where fluvial or wave erosion, coastal ice push, skin flows, or some other form of mass movement removes insulating

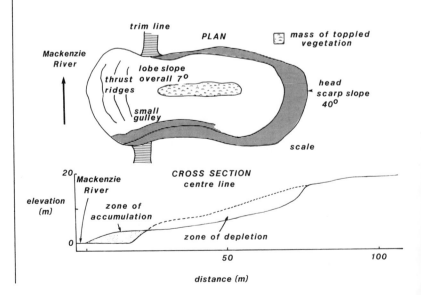

Figure 3.8.
The morphology of a bimodal flow near Hanna Island, NWT (McRoberts and Morgenstern 1974).

vegetation or unfrozen soil. This exposes ice-rich or thaw-susceptible soil, which melts and falls, slides, or flows down the headscarp into the muddy, bulging **zone of accumulation**. As such flows often terminate in water, undercutting of the toe can initiate further movement. Many bimodal flows are active in late fall, when their floors become fluid masses of mud. Active scarps retreat at rates ranging from 1.5 to 15 m per year. Bimodal flows are found along steep slopes adjacent to lakes, along the Arctic coast, and on the banks of rivers. They are common in fine-grained marine, lacustrine, and deltaic deposits on the Arctic Coastal Plain to the east and west of the Mackenzie Delta, and on the western Arctic islands; they are less common in the eastern Arctic. Bimodal flows have also been reported from the hills and mountains of the St Elias Mountains, where they provide the material for debris flows and the construction of large fans. Retreating bimodal flows also play an important role in forming dry or stream-occupied valleys in this area (Harris and Gustafson 1988).

Complex Movements

Many slope movements can actually be defined as complex, with different types of movement operating at different times or in different parts of the moving mass. They include earthflows that develop from slumps, and large and very dangerous debris avalanches, which begin as rockfalls or rock slides and are transformed into flows. Thus although it was useful to discuss the Frank, Hope, and Avalanche Lake slope movements under the heading 'debris avalanches', they had fall as well as flow components, and were therefore complex movements.

Subsidence

Subsidence is the predominantly vertical downward movement or sinking of superficial parts of the Earth's crust. Although it is included in this chapter for convenience, it is not a downslope or mass movement. Subsidence involves material that is confined on all sides, and though less widespread than slope movements, it can be more deep-seated. Fairly rapid subsidence is caused by the collapse of the roof of a cave or some other large cavity beneath the surface. Suitable cavities can be produced by mining, coastal erosion, piping (Chapter 8), subsurface solution (Chapter 10), melting of ground ice (Chapter 7), chemical changes involving reduction in volume, vulcanicity, or old land movements. A second type of subsidence involves more gradual settlement, owing to the reduction in volume of small cavities or pore spaces in the soil. Such settlement can result from weight being added to the surface, lowering of the groundwater level, or removal of oil or water at depth. Sea and lake bed deposits undergo widespread and extensive settling, but on land it is often associated with the loading of ice sheets, and man-made structures and fills.

4 | Glaciers and Ice Ages

Although the major elements of Canada's scenery are the result of its **tectonic** history and the nature of the underlying rocks (Chapter 1), the character of its landscapes reflects the way geomorphological detail has been superimposed onto the geological backcloth. Much of this detail has been provided by glaciation. Canada was the site of the first recorded ice age on Earth, and it supported the largest ice sheet in the northern hemisphere in the most recent **glacial** stage of the present ice age.

Glacial Budgets

An active glacier has two main components: an accumulation zone above the snowline, where more snow falls in winter than is lost in summer, and an **ablation** zone below the snowline, where more snow is lost in summer than falls in winter (Fig. 4.1). At the snowline or, more correctly, the equilibrium line, the snow and ice added in a budget year is exactly balanced by the amount that is lost. Ice loss, or ablation, occurs mainly through melting in temperate regions, but **calving**, or the breaking off of icebergs and smaller fragments into water, is the most important mechanism in Antarctica. Calving into ice-dammed lakes and the sea was also of some importance in Canada and elsewhere in the northern hemisphere during the latter part of the last glacial stage.

The slow transformation of freshly fallen snow into glacial ice involves the growth of ice crystals and the elimination of the air spaces between them. This is accomplished under pressure as the snow in the accumulation zone becomes more and more deeply buried in snow banks. Meltwater plays an important role in the formation of glacial ice in temperate latitudes, and it therefore takes much longer to form in very cold polar regions.

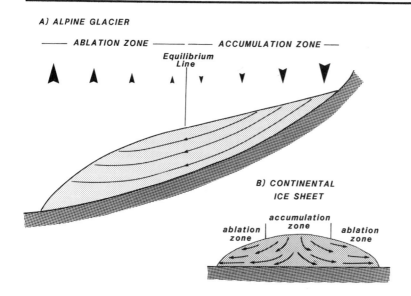

A) ALPINE GLACIER

——— ABLATION ZONE ——— ——— ACCUMULATION ZONE ———

Equilibrium
Line

B) CONTINENTAL
ICE SHEET

ablation | accumulation | ablation
zone | zone | zone

Figure 4.1.
Glacial accumulation-ablation zones and ice movement in valley glaciers and continental ice sheets. The size of the arrows above the alpine glacier refers to the relative amount of accumulation (pointing down) or ablation (pointing up) that takes place along its length.

Ice Temperature

The temperature of the ice at the bottom of a glacier plays a crucial role in determining how it erodes, transports, and deposits material. Ice temperature depends upon air temperature and solar radiation, the temperature of the snow that is being transformed into glacial ice, and the **latent heat** released by the refreezing of meltwater. Ice temperatures are also affected by geothermal heat coming from the Earth, and by heat generated by internal friction within the ice as it moves.

The temperature of cold ice is well below the **pressure melting point**, but the temperature of warm ice is close enough to that point to contain liquid water. Alpine or mountain glaciers in the middle latitudes generally contain warm ice. Cold ice is usually found in severe polar climates, although the ice may be thick enough in some areas to raise the temperature to the pressure melting point.

Ice Movement

An active glacier is a transportation system, carrying snow and ice from the accumulation zone to dispose of it in the ablation zone (Fig. 4.1). A glacier or, more strictly speaking, the glacial terminus or snout, can advance, retreat, or remain stationary, usually in response to changes in climate, but as long as there is an accumulation zone, the ice itself will continue to move towards the terminus.

A glacier moves because of stresses imposed by the weight of the overlying ice and the slope of the ice surface. Several mechanisms are involved:

(a) Ice flows or creeps as a result of the displacement or movement of individual ice crystals. The rate of creep is very sensitive to ice thickness, and fairly small increases in thickness cause much greater increases in the rate of ice creep. The flow rate is also dependent on ice temperature, with the highest rates occurring in warm ice.

(b) Ice fractures or breaks when it is unable to adjust to the applied stresses. This is unlikely to occur beneath very thick ice where creep is effective, but **crevasses** are **tensional** fractures on the ice surface, and **shear** fractures, which involve the movement of ice along slip planes, are important in the thin ice near the glacial terminus.

(c) Glaciers can also move by sliding or slipping over their beds. The relative and absolute importance of sliding depends upon such factors as bed slope and ice thickness and temperature. A glacier with cold ice at its base is frozen to its bed, and no sliding can take place. The presence of meltwater on the bed of a warm glacier probably helps it to slide, but it may be largely absent if the bedrock is particularly **permeable**. It is not yet known whether meltwater occurs as a thin film, or in cavities in the base of the ice.

In some areas, much of the forward movement of the ice may have been the result of the movement or deformation of its bed (Boulton et al. 1985, Fisher et al. 1985). Deformation of soft, wet **sediments** beneath the ice prevents it from steepening to attain a 'normal' parabolic profile. According to this theory, the ice would therefore have been thinner and flatter where it lay on the deformable sediments of the Interior Plains than where it lay on the hard, stable rocks of the Shield. Similar variations in ice steepness could reflect the presence of dry, stable material beneath cold ice and wet, deformable material beneath warm ice. Deformation of the glacial bed by thin ice could account for small amounts of **isostatic** recovery, high rates of ice retreat, and the presence of ice-thrust ridges indicating bed deformation (Chapter 6) along the southern and western margins of the Laurentide ice sheet.

Ice velocity and thickness increase from the head of a glacier down to the equilibrium line, then decrease down to the terminus. Ice flow is therefore accelerating or extending in the accumulation zone, and decelerating or **compressing** in the ablation zone. Extending flow may also occur where the bed steepens, the ice thins and becomes crevassed, and velocity increases by sliding. Compressive flow occurs where the bed becomes concave or flattens, the ice thickens, and velocity decreases. Suitable sites include places where valleys become narrower, at the foot of icefalls, and at the ice terminus. Where the flow is extending, stresses in the ice produce planes of weakness that curve downwards to the ice bed in a down-glacial direction. Where flow is compressive, however, the planes of weakness curve up to the ice surface in a down-glacial direction, helping to evacuate **debris** from the base of the ice and facilitating **erosion** of the rock bed (Fig. 4.2).

Glaciers may advance suddenly, attaining rates of flow ranging from

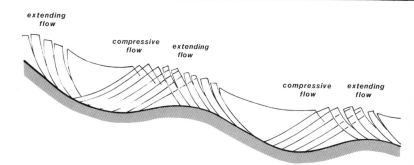

Figure 4.2.
Compressive and extending ice flow over an uneven bed.

about 150 m to more than 6 km per year. Surging glaciers are common in the St Elias Mountains in the Yukon, but they have been observed elsewhere, including on Ellesmere, Axel Heiberg, and Baffin Islands. It has also been suggested that surging may have occurred in the last glacial stage in the western Interior Plains and in the Hudson Bay and Great Lakes regions. Glaciers that surge seem to do so repeatedly, and probably at fairly regular intervals of from about 15 to more than 100 years. It is not known why surging occurs, although one possible cause that is often cited is an increase in the temperature of the ice at its base, possibly as a result of increasing ice thickness; another is the presence of a **basal** film of water of increasing thickness (NRC 1969).

Glacial Erosion

Glaciers use rock fragments or abrasives at their base to scrape the underlying bedrock. The abrasion rate initially increases with increasing ice thickness, until the pressure exerted on the abrasives makes them increasingly difficult to move. The abrasion rate then declines with increasing pressure until it becomes impossible to move the rock fragments, which are then deposited as lodgement **till** (Fig. 4.3). Abrasion also increases with the speed of the ice at its base. Ice velocity determines the number of abrasives that can be scraped along a glacial floor and the amount of debris that can be removed. The amount of water at the bed of a glacier is also important. The presence of water reduces friction and increases the sliding velocity, but if there is too much water it buoys up the base of the ice and reduces abrasion. Compressive flow, the local increase in ice thickness, and the build-up of water in depressions underlain by impermeable rock help to explain why glacial erosion tends to emphasize rather than reduce inherited irregularities in the landscape.

Other erosional processes quarry larger rock units from bedrock, although they are generally not well understood. One possible mechanism involves stresses exerted by glaciers against bedrock obstructions. The pressures exerted by moving ice against the upstream side of a rock obstruction would be higher than at the more sheltered downstream end (Fig. 4.4). It has been proposed that these differences in stress eventually cause the rock to fracture, loosening blocks that can then be removed by the ice.

Glacial erosion is most effective under warm-based ice, where the bed-

Figure 4.3.
The relationship between abrasion rate and ice thickness and velocity. Each line represents a different ice velocity (Boulton 1974).

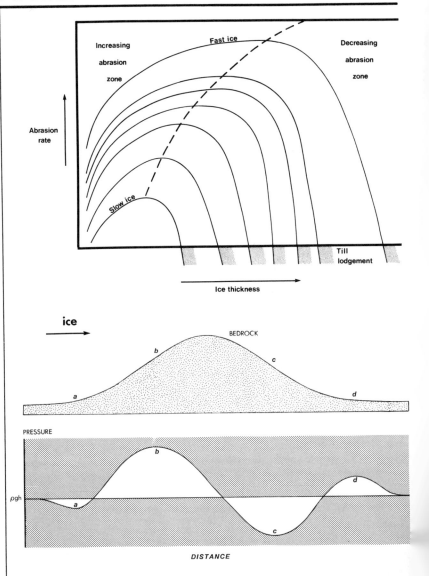

Figure 4.4
Pressures exerted by a glacier against a bedrock obstruction. ρgh is the pressure exerted on a horizontal surface by the ice, where ρ is ice density, g is gravity, and h is the ice thickness. Pressures are highest at b, where the ice is pressing against the upstream side of the obstruction, although there is a secondary zone of high pressure at d, where the ice presses against the foot of the downstream side. The low pressure at c is due to the ice moving away from the downstream flank of the obstruction, while bridging of the ice produces a secondary zone of low pressure at a (Boulton 1974).

rock can be abraded by debris-laden ice moving over its bed on a thin film of water (Fig. 4.5a). Abrasion is usually insignificant beneath cold-based ice that is frozen to its bed (Fig. 4.5b). Cold ice can abrade its bed, however, if it is supplied with debris from a zone of warmer ice up-glacier. This could occur where a thick central portion of an ice sheet is warm-based, while the thinner ice near the margins is cold-based. As ice moves from the warm- into the cold-based zone, water and debris freeze onto the base in successive layers as they are **thrust** up over the frozen basal ice. Abrasion can then occur in the cold-based marginal zone if the **entrained** debris extends down to the bedrock through the several centimetres of immobile ice (Fig. 4.5c). Topography can also play an important role: a cold-based

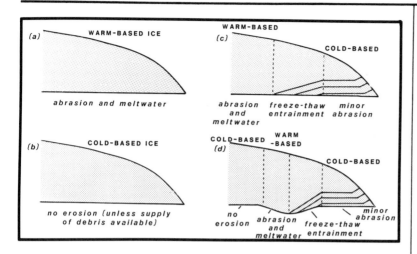

Figure 4.5.
The effect of ice temperature on erosional processes (Sugden and John 1976).

glacier can become warm-based where the ice is thicker over major depressions. This zone would then arm the cold, downstream ice with basal debris, allowing erosion to occur in the marginal zone (Fig. 4.5d).

The positions of zones of warm, cold, and transitional ice shift through a glacial stage as the climate changes. Areas that were once largely protected from glacial abrasion and atmospheric weathering by cold ice may later experience a period of more active erosion under warm ice. On the other hand, the presence of glacial erosional features in Antarctica, in areas where the ice is presently frozen to its floor, suggests that they were under warm-based ice at some time in the past.

Glacial Deposition

Sediment deposited by glacial ice is usually referred to as till, although the terms 'unstratified **drift**' and 'boulder clay' have also been used. Tills tend to be poorly sorted mixtures of sub-angular stones in a matrix of sand, silt, or clay, but the range of till types is enormous. Till is derived from debris obtained from the beds and sides of glaciers, and in some cases from material falling onto the ice surface from the sides of valleys or high mountain peaks. This material, which can be carried on (supraglacially), in (englacially), or under the ice (subglacially), can then be deposited in a variety of situations. One may therefore distinguish tills according to their **depositional** environment:

(a) Lodgement tills are deposited beneath glaciers, particularly if the ice is warm. These tills are formed either where small particles are freed from the ice by pressure melting and then plastered onto the bed, or where variable mixtures of ice and debris shear in layers at the ice base. An increase in the thickness of the ice or the roughness of the bed will tend to increase the rate of till deposition. Lodgement tills are laid down most thickly in depressions, and therefore tend to mask topographic irregularities.

(b) Subglacial flow tills are water-soaked sediments that are squeezed up into cavities in the base of the ice.

(c) Englacial debris is contained within the ice. This material is released and deposited as the ice slowly melts from above and below. There are likely to be large amounts of englacial material in the entrainment zone where warm ice passes into cold, and in the cold zone that lies downstream (Fig. 4.5 c, d). Melt-out tills are therefore more characteristic of the outer portion of glaciers, while lodgement tills are generally deposited further inside the ice.

(d) Supraglacial flow tills form near the ice terminus from supraglacial debris and englacial material melting out at the surface. The presence of so much meltwater at a sloping ice terminus usually results in the redistribution of the water-soaked slurry by sliding and flowing.

(e) Tills deposited at the front of an advancing glacier consist of pushed or bulldozed material and sediment brought to the terminus by normal glacial transport. Frozen ground that is bulldozed by a cold-based glacier is sheared, **faulted**, and folded. There is also some faulting and thrusting when warm ice slides over an unfrozen surface, but folding is usually more common, and flowage and **glaciofluvial** processes are active.

(f) Some tills are composed of sediments deposited by grounded or floating ice in glacial lakes or in the oceans. These sediments can be deposited at the foot of the ice terminus, or transported over considerable distances by icebergs. Some Canadian tills were deposited under water in the latter part of the last glacial stage, and they are still being laid down in Antarctica and in many Canadian fiords, although little is known of the precise modes of deposition.

Ice Ages

Scratched and moulded rock surfaces and tills containing **erratics** brought from distant sources bear testimony to the repeated glaciation of our planet through geological time. Evidence for these ancient ice ages has been found in a variety of areas, including some that have now become hot deserts (John 1979).

We know that there were several ice ages in the vast Precambrian Eon. The Huronian glaciation, about 2,300 million years ago, is the earliest ice age known on Earth. The best evidence is found on the north shore of Lake Huron in Canada, where the rocks suggest that there were at least three periods of glacial advance and retreat. There is also some indication of this or a somewhat later ice age in southeastern Africa and northwestern Australia. After an extremely long period lacking any trace of widespread glaciation, evidence of several ice ages appears on virtually every continent. These ice ages may have reached their peaks about 900-950, 750, and 600 million years ago.

There are traces in the Tropics of an Ordovician ice age culminating about 450 million years ago. It affected a large area on the old continent of **Gondwanaland**, in western and northern Africa and eastern South

America. Independent ice centres may have existed in Brazil, and in south-western, western, and northeastern Africa. Evidence in the Sahara Desert includes small roches moutonnées, striations, friction marks, fluted moraines, and sandstones formed from outwash deposits and eskers (Chapter 6).

The Permo-Carboniferous ice age reached its peak on the continent of Gondwanaland between 310 and 270 million years ago. As the supercontinent changed its position with respect to the south pole, ice centres shifted from South America and southwestern Africa to South Africa, Madagascar, and India, and still later to Antarctica and Australia. Southern Africa, however, was the main centre of glaciation for the longest period of time, and it provides the best evidence for this ice age.

Although most ice ages lasted between 20 and 50 million years, it has been suggested that they may have reached their peaks about every 150 million years. The lack of convincing evidence for an ice age about 150 million years ago could simply reflect the fact that there were no large land masses near the poles at that time. The distribution of the continents might also explain the absence of ice ages for long periods during the Precambrian. On the other hand, of course, any apparent degree of regularity in the occurrence of ice ages in the past could be purely coincidental.

The Late Cenozoic Ice Age

The present ice age began when large, stable ice sheets developed over Greenland and Antarctica (Table 4.1). Although we don't know when glaciation began in Greenland, deep-sea sediments to the south of the island show that debris was being deposited beneath floating ice about 3.5 million years ago. Antarctic glaciation may have begun as long as 38 million years ago, and the continent was ice-covered 20 to 25 million years ago. The growth of these enormous ice sheets must have helped to cool the Earth, providing conditions that, by the Pleistocene, were suitable for the development of continental ice sheets in the northern hemisphere.

It was not very long ago that the last ice age was considered to be synonymous with the Pleistocene epoch or, if we add the 'nonglacial' **Holocene** or Recent epoch of the last 10,000 years, with the **Quaternary** period (Tables 1.1 and 4.1). Moderately sized ice sheets began to form in the northern hemisphere between 2.55 and 2.4 million years ago. This was not the beginning of the present ice age, however, because there was a great deal of ice on our planet long before that time. Nevertheless, changes in climate became more frequent and rapid in the Pleistocene than they had been previously. This caused the alternate growth and decay of vast ice sheets in the northern hemisphere, migration of the world's climatic and biological zones, and changes in sea level of 100 m or more (Fig. 4.6).

The Pleistocene consisted of a series of glacial and interglacial stages. During the glacial stages, mean annual temperatures were as much as 10°C colder than today in the interior of some continents, and large areas of

Table 4.1   Major divisions of the Late Cenozoic Ice Age.

ERA	PERIOD	EPOCH	STAGE	SUBSTAGE	ISOTOPIC STAGE
0 Ma		RECENT (Holocene)			most of 1
		——— 10 ka ———			
			Wisconsin Glacial	late	part of 2 and 1
				——— 23 ka ———	
				middle	3 and part of 2
				——— 64 ka ———	
		late		early	4
				——— 75 ka ———	
CENOZOIC	QUATERNARY	PLEISTOCENE	Sangamon Interglacial		5
		——— 130 ka ———			6 to about stage 9
		middle			
		——— 790 ka ———			
		early			
	——— 1.65 Ma ———				
65 Ma	TERTIARY				

Antarctic glaciation, and therefore the late Cenozoic Ice Age, may have begun as long as 38 million years ago. Odd numbers are used for warmer interglacial isotopic stages (with the exception of the stage 3 interstadial), and even numbers for colder glacial stages. 'Ma' and 'ka' refer to millions and thousands of years, respectively.

Table 4.2   Classical Pleistocene terminology in ascending order of age. Glacial stages are inset and are in heavier type than the interglacial stages.

North America	European Alps	Northern Europe	British Isles
Wisconsin	**Wurm**	**Weichsel**	**Newer Drift**
Sangamon	R\W	Eemian	Ipswichian
Illinoian	**Riss**	**Saale**	**Gipping**
Yarmouth	M\R	Holstein	Hoxnian
Kansan	**Mindel**	**Elster**	**Lowestoft**
Aftonian	G\M	Cromerian	Cromerian
Nebraskan	**Gunz**		

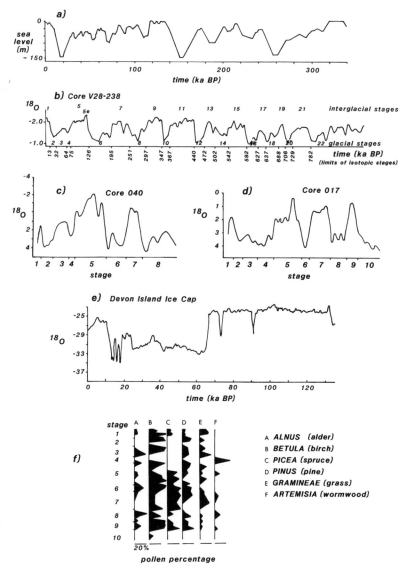

Figure 4.6.
Climatic indicators for portions of the middle and late Pleistocene: (a) sea level in New Guinea for the last 300,000 years (Chappell 1983); (b) isotopic deep sea core V28-238 (Shackleton and Opdyke 1973). With the exception of stage 3, which is now defined as an interstadial, odd stage numbers refer to interglacial stages and even numbers to glacial stages; (c) and (d) isotopic deep sea cores off Baffin Island (Aksu, in Andrews 1985); (e) isotopic core from the Devon Island Ice Cap (Koerner and Short, in Andrews 1985); and (f) proportion of pollen grains of different species in samples obtained from Baffin Bay. The types of vegetational community that existed at various times provide evidence of the type of climate at those times (after Mudie and Short, in Andrews 1985). (kaBP: thousands of years before present.)

North America, northern Europe, and western Siberia were covered by ice. The interglacial stages were intervening periods of warmer climate, when large ice sheets disintegrated and disappeared from the northern hemisphere. Although we are probably living today in an interglacial stage, about 11 per cent of the world's land area is still under glacial ice; however, only Antarctica and Greenland are presently covered by continental-scale ice sheets.

Until quite recently it was believed that the Pleistocene in Europe and North America consisted of four major glacial stages and three interglacials (Table 4.2). The glacials were thought to be rather rare, cataclysmic events, interrupting warmer interglacials of much greater duration. This view was based largely upon terrestrial evidence, which has had to be reassessed in the light of evidence obtained from sediments on the ocean floor.

Oxygen has three isotopes (atomic structures that vary according to the number of neutrons in the nucleus). O^{16}, the lighter and by far the most common isotope, evaporates a little more easily from sea water than do the heavier isotopes O^{17} and O^{18}. During a glacial stage some of this evaporated O^{16}-enriched water becomes glacial ice, resulting in slight enrichment of the sea with the heavier isotopes. The proportion of O^{18} to O^{16} therefore increases in sea water during a glacial stage and decreases during an interglacial (Fig. 4.7). Most workers believe that isotopic variations mainly reflect changes in the volume of the ice on Earth. Historical changes in the isotopic ratios are recorded in the shells of marine creatures that have accumulated over long periods of time among the sediments on the ocean floor. Oxygen isotopic analysis is usually done on microscopic foraminifera, although molluscs, coral, and other organisms have also been used.

There is broad agreement between climatic records obtained using a variety of techniques. These include oxygen isotopic analysis of deep-sea cores, speleothems (stalagmites and stalactites) in limestone caves (Chapter 10), and glacial ice in Antarctica, Greenland, and Arctic Canada; changes in sea level determined by the dating of tropical coral-reef shorelines; changes in the type and distribution of terrestrial and marine organisms; and pollen analysis of vegetational variations (Fig. 4.6).

Isotopic data suggest that there have been many glacial stages in the last two and a half million years. Each of the three traditional glacial stages preceding the **Wisconsin** must therefore represent several glacial episodes, while others occurred within the traditional interglacials. There may have been about forty cooler periods in the northern hemisphere between 2,500,000 and 900,000 years ago, occurring at intervals of about 41,000 years. The size of the ice sheets seems to have doubled in the last 900,000 years, however, and the glacial stages have increasingly occurred at intervals of about 100,000 years (Fig. 4.6b). Other cycles, of about 41,000 and 23,000 years, have been superimposed on the main glacial-interglacial cycles. Glacial cycles have been described as being sawtoothed, each long period of ice build-up during the glaciation phase being succeeded by a fairly short period of deglaciation.

There appears to have been roughly the same amount of ice on Earth in each of the glacial stages of the last 900,000 years, although the last (the Wisconsin in North America) may have been somewhat less severe. The glacial stages lasted about 100,000 years on average, compared with less than 20,000 years for the intervening interglacials. Conditions in many of the interglacial stages were similar to today's, but the last, which peaked about 125,000 years ago, was even warmer than at present. Sea level at this time was between 3 and 10 m higher than today, possibly because of the melting of the west Antarctic ice sheet.

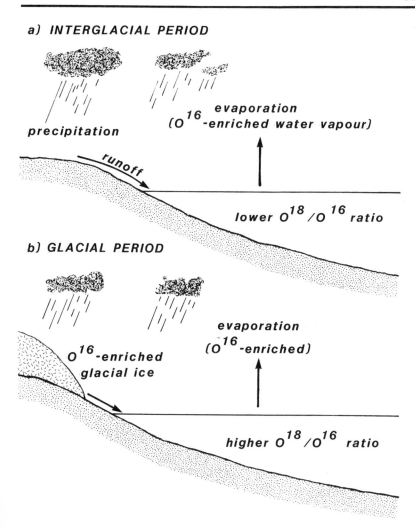

a) INTERGLACIAL PERIOD

precipitation

evaporation
$(O^{16}$-enriched water vapour$)$

runoff

lower O^{18}/O^{16} ratio

b) GLACIAL PERIOD

O^{16}-enriched
glacial ice

evaporation
$(O^{16}$-enriched$)$

higher O^{18}/O^{16} ratio

Figure 4.7
Oxygen isotopic variations in interglacial and glacial stages.

There were also marked fluctuations in temperature during the Wisconsin glacial stage, when **interstadials**, or intervals of higher temperature and ice retreat, alternated with **stadials**, or colder intervals of ice advance. Sea level and isotopic data generally suggest that the lowest temperatures were attained in the late Wisconsin, between about 15,000 and 20,000 years ago, although some terrestrial data, based upon the dating of glacial deposits and the distribution of **faunal** and **floral** species, indicate that temperatures were at least as low in the early Wisconsin, about 70,000 years ago. Temperatures rose fairly rapidly to their present level following the minimum in the late Wisconsin (Fig. 4.6).

The Origin of Ice Ages

Any attempt to explain the origin of ice ages must account for the very long nonglacial periods between them, and the shorter periods of fairly warm conditions within them. The mechanisms that generate ice ages, however, are likely to be different from those responsible for variations in climate within ice ages. The numerous theories that have been proposed may be classified according to whether they attribute climatic variations to internal changes in the Earth, or to external meteorological or astronomical factors.

Internal theories • Large-scale glaciation can occur only if there are large land masses in the middle to high latitudes, where the climate is cool and wet. The positions of the poles and the continents have shifted through geological time (Chapter 1). Continental drift may help to account for the Permo-Carboniferous ice age, but it is not clear to what degree, if any, it contributed to the late Cenozoic ice age. Movement of the land and the poles was certainly too small to play any significant role in the Pleistocene.

Glacial accumulation zones are created or enlarged when land rises above the snowline during mountain-building episodes. Tertiary uplift must have helped to trigger the present ice age. The European Alps, the Himalayas, and the Sierra Nevadas of California, for example, were uplifted by at least 2,000 m in the Pleistocene-Pliocene epochs. Other ice ages, including the Permo-Carboniferous, also seem to have occurred after periods of extensive mountain building.

Continental drifting and vertical earth movements also affect ocean currents, which transfer large quantities of heat about the globe. If the water over the ridge between Iceland and Norway became shallower, for instance, as a result of uplift or a drop in sea level, it would reduce the amount of warm, tropical water entering the Arctic. Land bridges created by uplift could help to initiate an ice age, but the mechanism is more likely to be significant once glaciation has caused a drop in sea level.

External theories • Mounting evidence suggests that cyclical variations in temperature within ice ages can be attributed to changes in the Earth's orbital geometry around the sun. Elements of the astronomical theory were suggested by Croll in 1875, and developed by Milankovitch in the 1920s. The amount of solar radiation reaching the upper limits of the atmosphere varies according to

(a) wobbling of the Earth's axis with a period of 19,000 and 23,000 years;

(b) changes in the angle made by the Earth's axis with the plane of orbit, with a period of 41,000 years; and

(c) changes in the shape of the orbit with periods of 95,000 to 136,000 and 413,000 years.

These cycles affect the seasonal and geographical distribution of solar radiation. Many workers have argued that the similarity between variations in solar radiation, sea level, and oxygen isotopic ratios is probably too great to be coincidental. Nevertheless, while orbital variations are probably the main cause of glacial-interglacial cycles, it is not yet clear how these small changes in radiation trigger major climatic fluctuations.

A number of meteorological mechanisms could have played a secondary role in perpetuating or terminating a glacial stage. Meteorological explanations are based upon the fact that the effects of fairly small changes in climate can be amplified in some critical situations. As a simple example, a small drop in temperature would be sufficient to freeze an ocean that was already near the freezing point. This would increase the surface **albedo** or reflectance of the sea to incoming solar radiation, further lowering the temperature. A similarly self-perpetuating system could be created by the chance occurrence of a series of particularly snowy winters and cool summers. This might allow snow to persist throughout the year, increasing the surface albedo sufficiently to create a cooler local climate, and ensuring the continued accumulation of snow. Other mechanisms, which are self-regulating, may help to explain both the onset and termination of glacial stages. For example, an ice-free Arctic Ocean provides moisture for the growth of glaciers on the surrounding mountains and plateaux. As the ice becomes more extensive, however, it promotes climatic cooling and the eventual freezing of the ocean. This then denies further moisture for glacial nourishment, causing the ice to waste away.

Climate can be affected by a change in the character of the atmosphere, including a change in the proportion of its gaseous constituents. Cause-and-effect relationships are very complex, however, and poorly understood. Many hypotheses involve the **greenhouse effect**, which depends upon the fact that incoming solar radiation is mainly shortwave, whereas outgoing radiation from the Earth is largely longwave. The atmosphere is warmed by the absorption of outgoing longwave radiation by carbon dioxide and water vapour. The amount of carbon dioxide in the air increases during a glacial stage because there is only a small amount in the ice sheets. This increases the absorption of outgoing radiation, heating the atmosphere and helping to bring glaciation to an end. On the other hand, ice ages could be the result of cooling caused by a reduction in the amount of atmospheric carbon dioxide, possibly following periods of major carbonate deposition in the oceans.

Suspended volcanic dust from eruptions filters out solar radiation, but it also increases the greenhouse effect. There was a marked increase in the number of eruptions in the Quaternary, and some workers believe that there has been a relationship between climatic variations and volcanic eruptions in the last few hundred years. But the data do not really allow any firm conclusions to be made, and in any case, the effect of eruptions is thought to be short-lived, as volcanic dust remains suspended in the atmosphere only for up to about seven years.

Several workers have attributed changes in climate to sunspot activity, but others argue that the relationship is insignificant. Sunspots influence the quality as well as the quantity of solar radiation reaching the Earth. Ultra-violet and x-ray radiation associated with solar flares affects the production of ozone. This could induce short-term climatic variations related to the sunspot cycles, which trigger or reinforce longer-term trends. Long-term variations in solar emission are of potentially greater importance, however, possibly over cycles of 200,000 and 400,000 years.

The Antarctic ice-surge theory envisages Antarctic ice as becoming thick enough to melt at its base, allowing it to surge into the surrounding ocean to form a vast ice shelf. The increased albedo of the smooth, white ice shelf, relative to the sea water that it covers, lowers global temperatures, allowing continental ice sheets to develop in the northern hemisphere. The glacial stage ends when the ice shelf is broken up by waves and tides, thereby lowering the albedo and promoting global warming. The next glacial stage occurs when the ice once again becomes thick enough to promote basal melting.

Conclusion • There is no simple answer to the question of what caused ice ages to occur, nor is there ever likely to be one. If ice ages are cyclical, then their occurrence must presumably reflect periodic changes in solar radiation, possibly associated with the cyclical swirling of our galaxy through space, or chemical changes in the atmosphere, oceans, or **lithosphere**. Other non-cyclical factors may have helped to trigger ice ages, including land elevation and the position and arrangement of the land masses. The cyclical occurrence of relatively warmer and colder periods during ice ages appears to reflect variations in the Earth's orbital geometry, although numerous other factors must have had a contributory role.

The Glaciation of Canada 5

Almost all of Canada was glaciated at some time during the Pleistocene, although only about 1 per cent is under ice today. The largest unglaciated region was in the western Yukon, east of the coastal mountains, but there may have been some smaller unglaciated areas along the Mackenzie River, in the western Arctic islands, the Albertan Foothills, the southwestern Interior Plains near the international border, and northeastern and eastern Canada.

There are many ice fields and alpine cirque and valley glaciers on the eastern Arctic islands, but very few in the central and western Arctic (Fig. 5.1). This may be because of the higher land in the eastern Arctic and higher snowfall near Baffin Bay and the Atlantic Ocean (Fig. 1.13). Isolated peaks or **nunataks** project through the surface of ice sheets on the mountains of Baffin, Bylot, Axel Heiberg, and Ellesmere Islands. There is little glacial ice on the mainland of eastern Canada; nevertheless, small pockets persist at the bottom of cirques in the Torngat Mountains of northern Labrador.

Some of the most impressive alpine glaciers in Canada flow to the Pacific coast from large ice fields in the St Elias Mountains of the Yukon (Fig. 1.9). Heavy snowfall from Pacific storms also nourishes ice fields and valley glaciers on the western slopes of the Coast Mountains of British Columbia. There are generally far fewer glaciers on the drier eastern slopes and inland ranges of western Canada, although they are fairly common in the Selkirk Range and in parts of the Rocky Mountains, particularly in the Columbia Icefield and in the cirques and valleys of the Mount Robson Massif (Fig. 5.1).

Figure 5.1.
Glaciers in Canada today.

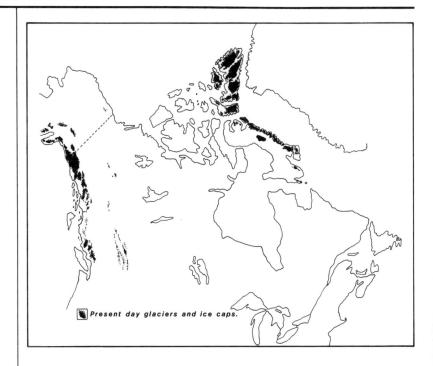

Present day glaciers and ice caps.

The Older Glacial Record

In most of Canada, evidence of earlier glaciations has been largely obliterated or masked by the fresh landforms and deposits of the last **glacial** stage, the late or classical **Wisconsin** (Table 4.1). Nevertheless, although there have been only a few detailed studies of sediments older than the Sangamon or last **interglacial** stage, there is some evidence of glacial and nonglacial stages dating back to the early Pleistocene (Fulton 1984, Fulton 1989).

Pre-Sangamon Events

About 2 million years ago, a Laurentide ice sheet deposited Precambrian **erratics** from the Shield in the Wellsch area of southwestern Saskatchewan. This glacial event, the oldest recognized in Canada (in the present ice age), did not extend as far west as Medicine Hat, Alberta. The later Dunmore Glaciation, which reached the eastern slopes of the Rocky Mountains, was therefore the first Laurentide ice sheet to cover much of southern Alberta, and probably the most extensive glaciation to affect the southern portion of the Interior Plains. **Tills** and other glacial deposits are sandwiched between flat-lying **basalts** near the middle Fraser River. These sediments, which are about 1.2 million years old, are evidence of the first known regional glaciation in south-central British Columbia. Remnants of tills at Quesnel in central British Columbia, near the centre of the former Cordilleran ice sheet, record the occurrence of as many as three pre-Wisconsin glaciations (Clague 1988). There is further evidence of pre-Wisconsin glaciation in several other areas. In the Hudson Bay Lowlands, the presence of tills beneath deposits that are thought to be of Sangamon age suggests that there were at least four glacial advances in this area before the last interglacial. On Banks Island in the western Arctic, glacial deposits are separated by interglacial sediments and underlain by nonglacial material. These deposits show that there were at least two glacial stages in this area before the Wisconsin Amundsen Glaciation: the Banks (more than 730,000 years ago) and the later Thomsen. There are only isolated deposits of probable pre-Sangamon age in Atlantic Canada. Among them is the Bridge-water Conglomerate, an iron-cemented till and outwash deposit along the coast of Nova Scotia that appears, on the basis of deep weathering, to have experienced at least one long interglacial stage.

The Sangamon Interglacial

The Don Beds in the Toronto area consist of up to 8 m of well-**stratified** clay and sand deposited in a **lacustrine** environment that was probably close to a large river. The organic remains of plants and animals that do not live this far north today show that for several thousand years the local climate was as much as 2 to 2.5 °C warmer than at present. The Don Beds are therefore usually assigned to the Sangamon interglacial, with the under-lying York Till representing the preceding glacial stage (Fig. 5.2). The depth of the water increased from about 2 to 20 m during the **deposition** of the Beds, and they are overlain by a regionally extensive deltaic deposit, which indicates further deepening and decreasing temperatures (Eyles and Clark 1988).

Sediments with plant and animal remains of probable Sangamon age have been identified throughout the western Interior Plains, and on the

Figure 5.2.
The late Pleistocene stra-tigraphy of the Toronto area. The exposure in the classical site, the Don Brickyard, is shown in the box (Eyles and Clark 1988). The term 'diamict' is defined in the glossary.

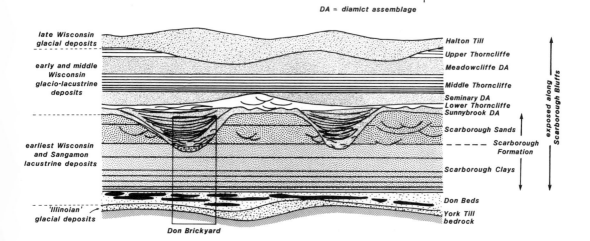

DA = diamict assemblage

late Wisconsin glacial deposits — Halton Till — Upper Thorncliffe — Meadowcliffe DA — early and middle Wisconsin glacio-lacustrine deposits — Middle Thorncliffe — Seminary DA — Lower Thorncliffe — Sunnybrook DA — Scarborough Sands — earliest Wisconsin and Sangamon lacustrine deposits — Scarborough Formation — Scarborough Clays — Don Beds — York Till — bedrock — 'Illinoian' glacial deposits — Don Brickyard — exposed along Scarborough Bluffs

Boothia Peninsula and southern Banks Island in the Arctic. The Missinaibi Beds in the Hudson Bay Lowlands are also generally considered to correspond to the last interglacial stage. The last interglacial is thought to be represented in Nova Scotia by organic beds with fossils indicating a warmer climate than today's. There is also at least one widespread wave-cut shore platform in Atlantic Canada, together with associated nonglacial deposits, 2 to 6 m above present sea level. This platform may be the eastern Canadian counterpart of globally extensive terraces cut during the last interglacial, when sea level was several metres higher than it is now.

There is some uncertainty over the definition of the Sangamon, or last interglacial stage, in Canada, depending in part upon whether local or global criteria should take precedence. If one restricts the Sangamon to isotopic substage 5e (Fig. 4.6b), the warmest part of the last interglacial, then the Wisconsin glacial stage would have begun about 116,000 years ago. Alternatively, if one accepts the broad global definition, then the last interglacial corresponds to the whole of stage 5 in the oxygen isotopic record (Table 4.1). There may have been substantial amounts of ice in southern Canada, however, during at least the latter part of stage 5. As noted above, the elevation of the Don Beds above Lake Ontario suggests that the water level was as much as 20 m higher towards the end of the last interglacial than it is today. Possible nonglacial explanations for this high lake level include postglacial **isostatic** recovery of the eastern outlet of Lake Ontario during the Sangamon, sediment- or water-loading of the lake basin, or subsequent lowering of the outlet by glacial **erosion** during the Wisconsin. Alternatively, there could be a glacial explanation involving the isostatic depression of the lake basin by the growing Laurentide ice sheet, or the blockage of its outlet through the lower St Lawrence by ice.

Using the global stage 5 definition of the last interglacial stage would place the Bécancour Till, deposited during an early advance of the ice into southern Québec, in the last interglacial. The last interglacial would therefore have begun about 130,000 years ago, and ended after the deposition of the nonglacial lacustrine and fluvial St Pierre sediments, which have been tentatively dated at about 75,000 years. These sediments, which overlie the Bécancour Till between Sorel and Québec City, represent a period 2 to 4 °C cooler than today, when ice was absent from the St Lawrence Valley. Nevertheless, it should be noted that these age assignments are the subject of considerable controversy, and in the absence of reliable dating, it could be argued that the St Pierre sediments are as young as middle Wisconsin or as old as Sangamon interglacial substage 5e (Vincent and Prest 1987).

Early and Middle Wisconsin Events

On the assumption that the last interglacial is represented by the whole of isotopic stage 5, the last glacial stage can be subdivided into early (about 75,000 to 64,000 years ago), middle (64,000 to 23,000 years ago), and late (23,000 to 10,000 years ago) Wisconsin substages. We have a fairly detailed record of the glacial history of the last 20,000 years, but because of the limitations of the **radiocarbon** dating technique, we know far less about the period between 20,000 and 40,000 years ago, and our knowledge is only very sketchy on the earlier portions of the Wisconsin glacial stage and the preceding Sangamon interglacial. The relative age of many deposits beyond the range of radiocarbon methods is commonly determined by attributing warm climate, organic, nonglacial deposits to the Sangamon or last interglacial stage. Nonglacial materials lying between tills above the assumed interglacial beds are then assigned to the middle Wisconsin, especially if they contain organic remains that indicate a climate cooler than today's. The possibility of error based upon these assumptions is clearly enormous, however, and until new dating techniques are developed, theories on the form, development, and extent of the ice in the early and middle Wisconsin must be regarded as essentially speculative.

There is considerable disagreement over the extent of the ice in the early Wisconsin, and particularly in comparison with the late Wisconsin. The available evidence can be interpreted in a number of ways. Some have concluded that there was no ice in the early Wisconsin; others, that the ice was more extensive at that time than during the late Wisconsin (Vincent and Prest 1987). Many workers subscribe to the latter position, attributing

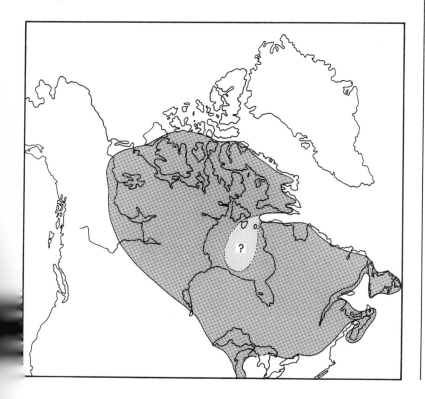

Figure 5.3.
Possible extent of middle Wisconsin ice with only moderate deglaciation. The lighter tone represents a possible marine incursion into Hudson Bay (Dredge and Thorleifson 1987).

Figure 5.4.
Possible extent of ice in the middle Wisconsin with considerable deglaciation (Andrews 1987).

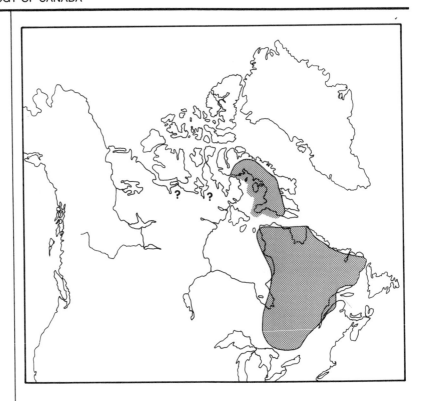

glacial deposits around the periphery of the area covered by the Laurentide ice sheet to the early Wisconsin. The evidence is not conclusive, however, because of the uncertainty over the age of these deposits, and many have now been reinterpreted and reassigned to pre-Sangamon glacial stages or to the middle or late Wisconsin.

There is also no consensus on the extent of the ice in the middle Wisconsin (Dredge and Thorleifson 1987). Some workers believe that there was only partial retreat of the ice margins during this period, while others think that the very core of the Laurentide ice sheet around Hudson Bay was free of ice (Figs. 5.3 and 5.4).

Evidence suggests that there was substantial retreat of the ice margins in southern Ontario in three middle Wisconsin **interstadials**. As the intervals between them were fairly short, however, they may be considered to have constituted a single long, generally ice-free period, possible interrupted by minor glacial re-advances. Evidence for the two earliest interstadials has been found at Port Talbot on the north shore of Lake Erie. The Port Talbot I interstadial may be about 65,000 to 50,000 years old, although it has not been accurately dated. It appears to have been warmer than the Port Talbot II (between about 47,000 and 40,000 years ago), but cooler than an interglacial. The Plum Point interstadial, between about 25,000 and 30,000 years ago, was a period of very cool climate. Similar, presumably middle Wisconsin interstadial deposits are scattered across southern Ontario, including the various units of the Thorncliffe Formation at Toronto (Fig. 5.2).

The west also experienced nonglacial conditions during the middle Wisconsin. There is no evidence of ice in the western Interior Plains during the middle Wisconsin Watino interval, which began before 52,000 years ago, and ended less than 23,700 years ago. The presence of nonglacial sediments throughout the Cordilleran region also suggests that there was substantial retreat or complete disintegration of the Cordilleran ice sheet in the middle Wisconsin. Glaciers were restricted to the mountains in southern British Columbia during the Olympia nonglacial interval or interstadial, which began before 59,000 years ago and lasted until about 20,000 to 25,000 years ago.

There are two or three till sheets above, and therefore younger than, the Sangamon Missinaibi Beds of the Hudson Bay lowlands. It has been suggested that the presence of water-laid deposits between these tills shows that Hudson Bay was at least partially ice-free in the Wisconsin, possibly about 75,000 and 35,000 years ago (Andrews et al. 1983). Others, however, have proposed that these water-laid sediments were deposited under the ice, and therefore do not represent nonglacial or ice-free conditions.

Ice appears to have persisted in some parts of southeastern Canada in the middle Wisconsin. The Montreal region and a portion of southeastern Québec and Anticosti Island were ice-free for part of the middle Wisconsin, but an ice cover probably remained north of the St Lawrence, on most of Newfoundland, and on the uplands of the Appalachians and Atlantic Canada. Ice may also have persisted in the middle Wisconsin in the Foxe Basin, and probably in northern Keewatin.

The Late Wisconsin

The continental ice sheet in the late Wisconsin had three main components:
  (a) the Laurentide ice sheet;
  (b) the glacial complex in the High Arctic; and
  (c) the Cordilleran glacial complex.

The Laurentide Ice Sheet

When temperatures began to fall at the beginning of the last glacial stage, snow and ice started to accumulate in northern Canada, forming ice caps that helped to further cool the local climate. These ice caps coalesced and the growing ice sheet advanced towards the moisture-carrying southwesterly winds. Most of the country was eventually covered by Laurentide ice between the eastern margins of the Rockies, the Parry Channel in the Arctic, and the eastern seaboard of Nova Scotia (Fig. 5.5).

The formation of large ice sheets in northern Canada would have required much more snow than the fairly small amount that falls today

Figure 5.5.
Ice in Canada in the late Wisconsin. The minimum and maximum limits represent the range of opinions held by workers on the amount of ice existing at that time (Prest 1984).

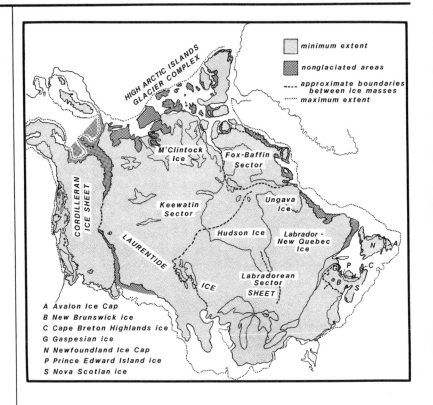

(Fig. 1.13). This may have been provided by a westward deflection of the West Greenland Current. As more warm water was carried northwards, an increase in the amount of open water in Baffin Bay would have resulted in an increase in the snowfall over Baffin Island and adjacent areas of northern Canada.

Flint (1943) believed that at the maximum extent of the late Wisconsin, ice flowed **radially** outwards from a central dome over Hudson Bay. Although the presence of thick ice beneath this dome appeared to account for rapid isostatic recovery in this area, there is no evidence of any westward flow of ice from Hudson Bay to Keewatin, and virtually none of eastward flow into Labrador-New Québec. Streamlined landforms, erratics, and other ice-flow indicators (Chapter 6) are generally dispersed in radial patterns around Keewatin and Labrador-New Québec (Fig. 5.6). It has been argued that these patterns developed only towards the end of the last glacial stage, after the sea had split the Hudson dome into eastern and western sections. It is difficult to account for the great distances some of the erratics travelled from their source areas, however, unless one accepts that the Keewatin and Labrador-New Québec ice centres were in existence for a considerable period of time (Shilts et al. 1979, Shilts 1980, Hillaire-Marcel et al. 1980).

The Laurentide ice sheet was much more complex and dynamic than

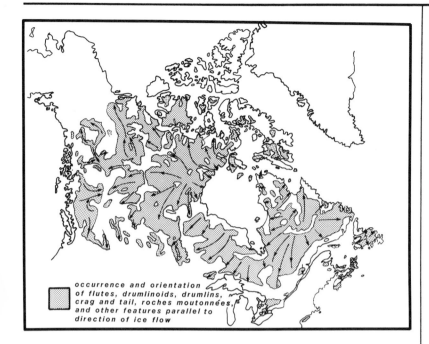

Figure 5.6.
The orientation of stream-lined features.

occurrence and orientation
of flutes, drumlinoids, drumlins,
crag and tail, roches moutonnées,
and other features parallel to
direction of ice flow

was once believed. Although some workers still support the single-dome concept, most authorities are now convinced that there were several centres of ice accumulation and dispersal, although they did not all exist at the same time. The multi-dome concept was first enunciated in 1898, when Tyrrell suggested that there were three main centres of outflowing ice, in Keewatin, Labrador-Ungava, and northern Ontario (the Patrician). The number and location of the Laurentide ice centres, however, are still problematic. Most workers recognize that there were at least three major sectors of outflowing ice, Labrador-New Québec (Labrador-Ungava), Keewatin, and Foxe-Baffin (Fig. 5.5). These sectors, which were in contact with each other for much of the late Wisconsin, maintained independent flow patterns controlled by their central domes and the ice divides that radiated from them. The different opinions that have been expressed on the precise form of the Laurentide ice sheet in the late Wisconsin are partly the result of the fact that the relative importance and position of these ice centres, divides with low points or saddles, and zones of confluence changed through time (Fig. 5.7).

Numerous ice divides have been proposed to account for the orientation of bedforms and the distribution of erratics in Canada. The M'Clintock ice divide, for example, separated ice flowing eastwards over the Boothia Peninsula and northern Keewatin from ice flowing westwards towards the Beaufort Sea (Dyke et al. 1982). Similarly, the Hudson divide separated ice flowing southwestwards over Ontario and northern Manitoba from ice flowing northwards over northern Hudson Bay (Dyke and Prest 1987). Fast-flowing ice streams, as opposed to sheet or normal regional ice flow,

Figure 5.7.
*Stages in the retreat of the
Laurentide ice sheet in the
late Wisconsin (Dyke and
Prest 1987).*

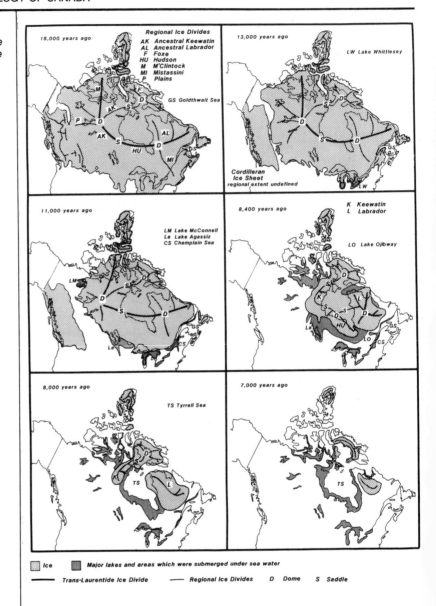

developed where there was ice convergence, some distance from saddles
or the junction of ice divides. They may have been helped by the presence
of fine carbonate till on the Shield of northern Ontario, which provided
wet, deformable beds for fast-moving ice (Hicock et al. 1989). Ice streams
may also have developed where **basal** ice moved off the Shield and onto
the deformable sediments of the Interior Plains (Chapter 4) (Fisher et al.
1985).

There continues to be considerable debate over the extent of the Laurentide ice sheet in the late Wisconsin (Fig. 5.5). Without reliable dating, it is difficult to determine whether glacial deposits, which show where the ice was, and nonglacial deposits, which show where it was not, should be assigned to the early, middle, or late Wisconsin, or even to older glacial or interglacial stages. As new evidence becomes available, deposits are reinterpreted and given new age assignments, necessitating periodic reassessment of the extent of the last ice sheet. The trend at the moment is to extend the limits of the late Wisconsin ice sheet from the more restrictive limits favoured by many workers in the early 1980s. While there is evidence of several glacial events in southern Alberta, for example, recent work suggests that the late Wisconsin was the only time Laurentide ice covered the west-central part of the province (Liverman et al. 1989).

Views are still rapidly changing on the palaeogeography of the Laurentide ice sheet in the late Wisconsin, and no conclusive reconstruction is yet possible. The following discussion is largely based upon a recent attempt to model the form and decay of the Laurentide ice sheet (Fig. 5.7) (Dyke and Prest 1987), but other interpretations of the available evidence are possible; many of the conclusions are tentative, and the subject of continuing debate.

Late Wisconsin palaeogeography • With the possible exception of the eastern Canadian Arctic, where the maximum recorded glacial advances took place between 11,000 and 8,000 years ago (Andrews and Miller 1984), the Laurentide ice sheet had attained its maximum dimensions almost everywhere by about 18,000 years ago. Laurentide ice was in contact with Cordilleran ice in the west, Appalachian and possibly Newfoundland ice in the southeast, and High Arctic ice in the north. The evidence for an ice-free corridor between the southwestern Laurentide ice sheet and the Cordilleran ice sheet during the late Wisconsin maximum is inconclusive (Fig. 5.5).

Ice flow was controlled by the L-shaped Trans-Laurentide superdivide, which connected the Keewatin and Labrador-New Québec sectors, and by other long regional divides radiating from the three ice domes (Fig. 5.7). There were also at least eleven major fast-flowing ice streams operating at this time. The largest may have occupied Hudson Strait, evacuating the enormous amounts of ice converging on northern Hudson Bay from the Keewatin, Labrador-New Québec, and Foxe-Baffin sectors. A southward-flowing ice stream also developed in Alberta and Montana at the confluence of Cordilleran-Laurentide ice, and others flowed eastwards and northeastwards along the St Lawrence Estuary and the Bay of Chaleur at the confluence of Laurentide and Appalachian ice.

There is longstanding disagreement between workers in New England, who believe that the northern Appalachians were dominated by Laurentide ice at the maximum of the late Wisconsin, and others who believe that the area was largely affected by local ice caps (Grant 1977). A number of ice caps coalesced, but their spread was restricted by the topography and the presence of deep submarine channels. Some coastal regions remained ice-free in the late Wisconsin, and some uplands in Gaspé and Newfoundland were probably never glaciated. There were active ice caps in the interior of Newfoundland and on the Avalon Peninsula. Laurentide ice may have reached the northern part of the northern peninsula of Newfoundland, although several workers believe that it crossed to the island only in the early Wisconsin, when Laurentide or Appalachian ice dominated Atlantic Canada.

In most areas there were only slight changes in the position of the ice margins from about 18,000 to 14,000 years ago. The ice front had retreated into the basins of Lake Erie and Lake Michigan by the latter part of this period, however, permanently deglaciating a central portion of southwestern Ontario (the Ontario Island) (Fig. 6.4). An 800 km re-advance of the James and Des Moines lobes into South Dakota and Iowa at the end of this period extended the local ice margins beyond their earlier limit. This rapid but short-lived surge represents the largest known oscillation of the Laurentide ice margins in the late Wisconsin.

Slow retreat over the next one thousand years separated the Laurentide ice sheet from the Cordilleran, and led to the development and growth of a series of ice-dammed lakes around its southern and western margins. The margins of the Laurentide ice sheet and the Newfoundland ice cap were still fairly close to their maximum positions, but the Appalachian ice complex survived only as remnant ice caps in Nova Scotia and Prince Edward Island.

The ice margins retreated much more rapidly after about 13,000 years ago, particularly in the Keewatin sector. The M'Clintock divide migrated to the east, and other divides were shortened or eliminated. On the other hand, there is evidence of an ice advance and retreat in eastern Baffin Island and on Banks, Victoria, and Melville Islands between 11,000 and 10,000 years ago. This resulted in the formation of the Hall Moraine in the Frobisher Bay area of southeastern Baffin Island shortly before 10,700 years ago. Although ice retreat in the west was more rapid than in southern and eastern Canada, the Newfoundland ice cap had become much smaller by 11,000 years ago and had broken up into separate ice masses. Withdrawal of the ice from the lower St Lawrence Valley allowed the Champlain Sea to occupy the lowlands of the Ottawa and St Lawrence Valleys between about 12,400 and 9,300 years ago (Figs. 3.5 and 5.8) (Parent and Occhietti 1988, Gadd 1989).

By 10,000 years ago the Cordilleran glaciers were not much larger than they are today. The Appalachian ice complex had disappeared, but the Newfoundland ice cap may have persisted as five small remnants. The southern margin of Labrador-New Québec ice was at the Québec North Shore Moraine at this time, marking the last major moraine-building episode in that area (Fig. 6.2). Continued ice retreat had caused substantial migration of the Hudson and Keewatin ice divides in the preceding one thousand years. The saddle or low point between the Hudson and Ancestral Keewatin ice divides intensified as ice flow converged more strongly on the **re-entrant** in the ice margin occupied by Lake Agassiz. This zone became the site of the Burntwood-Knife interlobate moraine. Another saddle developed on the eastern side of the Hudson divide, as ice converged on the re-entrant occupied by Lake Barlow. This boundary between Hudson and Labrador-New Québec ice became the site of the Harricana interlobate moraine. These developments resulted in the destruction of the Trans-Laurentide divide and increased the independence of the regional ice masses.

The Cockburn advance of Foxe-Baffin ice formed an extensive, though discontinuous, moraine system in northeastern Baffin Island between about 9,000 and 8,000 years ago (Andrews and Miller 1984). These moraines may mark the maximum position of late Wisconsin ice in eastern Baffin Island north of Cumberland Sound. Moraines of Cockburn age can be traced into southwestern Melville Island, and they are approximately contemporary with a series of moraines constructed along the northern flank of Keewatin ice (Chapter 6). The southern margin of Hudson ice at this time was surging repeatedly into Lakes Ojibway and Agassiz. The Cochrane ice surges, for example, were rapid southeasterly ice advances from Hudson Bay, between about 8,200 and 8,000 years ago. Ice advanced 200-400 km into Lake Ojibway and the James Bay lowlands on three occasions, resulting in the deposition of the Cochrane tills. Surging helped to thin the ice, allowing the sea to break into Hudson Bay along a narrow front from the north; this created the Tyrrell Sea. The rapid disintegration of the ice in Hudson Bay was further assisted by the northerly extension of Lake Ojibway along the zone of confluence between Hudson and Labrador-New Québec ice. There was a dramatic change in the character of the Laurentide ice sheet about 8,000 years ago, when the surviving ice dam separating the Tyrrell Sea from Lake Ojibway was breached by **calving**, and within a few hundred years, Hudson Bay was free of ice (Hillaire-Marcel and Occhietti 1980). This happened only about 75 years after the last of the three Cochrane surges. More rapid retreat of the Keewatin ice mass over several thousand years had left it by far the smallest of the remaining ice masses, and it had probably completely disappeared by 7,800 years ago or shortly thereafter. On the other hand, final retreat and disintegration of Labrador-New Québec ice, which at that time was by far the largest of the remaining ice masses, did not occur until about 6,500 to 6,000 years ago.

A similar marine invasion of the ice in the Foxe Basin did not take place until about one thousand years after it had occurred in Hudson Bay. By about 6,800 years ago, however, Foxe-Baffin ice remained only as large remnants on Baffin and Southampton Islands, and as a small ice cap on Melville Peninsula. By 5,000 years ago, the only remnants of Laurentide ice were on Baffin Island. Ice persisted in two ice caps, the Penny and Barnes, which probably separated from each other about 5,500 years ago. Each of these remnants has an area of about 6,000 km² today. They continue to retreat, although their destruction may not be completed before the onset of the next glacial stage.

The High Arctic Islands

The northern limit of Laurentide ice in the late Wisconsin was no further north than Bathurst Island and the southern part of Melville Island. Apart from fairly rapid postglacial rebound of the land, there is little obvious evidence of glaciation over much of the area to the north–possibly because the ice caps in this region were cold-based, in which case they would have left little trace of their existence (Chapter 4). Although it has been shown that some areas were certainly ice-free, the degree of ice cover in the High Arctic islands remains problematic. It has been proposed that the islands and the channels between them were covered by an Innuitan Ice Sheet that would have been contiguous with Laurentide and Greenland ice. More recent work, however, suggests that glaciation was more limited in the eastern High Arctic, and that the glacial sediments that do exist were deposited during an earlier glacial stage (England 1976).

The Cordilleran Glacial Complex

Nearly all of the Cordillera was glaciated during the late Wisconsin (the Fraser glaciation of the southern Cordilleran region), with the exception of some high mountain peaks and the more extensive unglaciated area in the far north (Fig. 5.5).

Intermontane, piedmont, and valley glaciers in the western Cordillera were fed by moisture-rich air masses from the Pacific. Snowfall was heaviest on the Coast Mountains and on the western slopes of the high mountain peaks. Glaciers from the coastal ranges flowed into the Pacific, and independent ice caps probably developed on the Queen Charlotte Islands and Vancouver Island. The ice cap on Vancouver Island was eventually overrun in the north by mainland ice, but there was only brief coalescence of mainland ice with the eastern portion of the ice cap on the Queen Charlotte Islands.

The Interior Plateau of British Columbia was gradually filled by ice flowing off the mountains to the east and west. The southern portion of this great ice sheet attained its maximum extent about 15,000 to 14,000 years ago, several thousand years after the Laurentide ice sheet. Ice flowed towards the dry Yukon Plateau to the north from areas of high precipitation in the St Elias, Coast, Cassiar, Selwyn, and Pelly Mountains, but it did not reach the limits attained by pre-Wisconsin ice sheets. There is still some debate over whether the ice became thick enough in the central portion of the Interior Plateau to produce a radially flowing dome with a surface higher than the surrounding mountain peaks, or whether this area was simply occupied by coalescent piedmont ice flowing from the mountains.

Glacial Lakes

About 56 per cent of the area once occupied by the Laurentide ice sheet was, or is still, covered by water (Fig. 5.8). As the ice retreated, the sea flooded isostatically depressed coastal regions, forming the Goldthwait Sea in the Gulf of St Lawrence east of Québec, the Champlain Sea in the St Lawrence Lowlands west of Québec, and the Tyrrell Sea in and around Hudson Bay (Chapter 9) (Fig. 5.7). Large areas of the Arctic and other smaller lowland areas around the Canadian coast also experienced marine **transgressions**.

There were long and narrow glacial lakes in the valleys of the western mountains, in southern and central British Columbia, in the middle Stikine Valley, and in the southwestern Yukon. The largest lakes, however, formed around the margins of the Laurentide ice sheet. Huge areas were flooded in western and southern Canada, where meltwater was confined between the ice margin and the higher land to the west and south, and the retreating ice sheet prevented natural drainage down to Hudson Bay. The large glacial lakes had a complex history of changing levels and sizes, as outlets opened and closed in response to isostatic uplift and fluctuations in the positions of the ice margins (Teller 1987).

The former lake bottoms are extremely level surfaces covered by clay

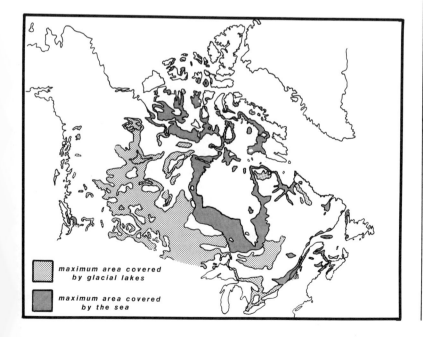

Figure 5.8.
Areas covered by glacial lakes or the sea during late Wisconsin and Holocene deglaciation.

maximum area covered
by glacial lakes

maximum area covered
by the sea

and silt. The position of former shorelines and nearshore areas may be marked by sand and gravel beaches, wave-cut terraces and bluffs, **lag** boulders, and deltas. Many deep valleys, occupied today by small underfit streams, were glacial spillways that carried the outflow from the lakes.

The Great Lakes

Although the relative contributions of the erosive mechanisms have not been determined, the Great Lakes were probably river valleys that were later scoured by glacial action. Conclusive evidence is lacking, but it seems likely that the basins had almost attained their present shape by the early Wisconsin, and possibly long before. The lake basins are partly geologically controlled. There are, for example, resistant Silurian Niagaran dolomites along the western and northern shores of Lake Michigan, and in the islands along the southern shore of the North Channel and the western shore of Georgian Bay in Lake Huron. Parts of the Michigan, Huron, and Erie Basins have developed in weak Devonian shales, and Lake Superior occupies a basin eroded in fairly weak Proterozoic rocks (Fig. 1.4).

The margins of the retreating Laurentide ice sheet became lobate in the Great Lakes Region, as the underlying topography influenced the form of the thinning ice. Ice retreat had uncovered the western end of the Erie Basin and the southern tip of the Michigan Basin by about 14,000 years ago (Fig. 5.9a). Lake Maumee, which developed in the Erie Basin, fluctuated between 232 and 244 m above sea level (asl), according to the position of the ice margin. Lake level was higher when drainage to the Mississippi was through the Fort Wayne-Wabash River system, and lower when it was through the Grand River into Lake Chicago. As the ice retreated further, the Erie, southern Huron, and Saginaw Basins were occupied by Lake Arkona (212-216 m asl), which drained through the Grand River spillway into the Michigan Basin.

About 13,200 to 13,000 years ago, the Port Huron (Mankato) ice advance separated Lake Saginaw from Lake Whittlesey (225 m asl) in the Erie Basin. Lake Whittlesey discharged into Lake Saginaw through the Ubly River channel, and then through the Grand River Valley into Lake Chicago and the Mississippi (Fig. 5.9b).

Ice margins fluctuated between 12,900 and 12,400 years ago (Fig. 5.9c). Lakes Warren (205-210 m asl) and Wayne (200 m asl) occupied the Erie Basin and the southern Huron and Saginaw Basins in the early part of this period. Drainage was through the Grand River Valley into Lake Chicago when lake levels were high, and eastwards when they were low. Ice retreat eventually caused the water levels to be lowered further, and between about 12,500 and 12,400 years ago the Erie, southern Huron, and Saginaw Basins were occupied by Lakes Grassmere and Lundy (195 and 189 m asl). These low level lakes may have discharged along an ice marginal stream across the northern peninsula of Michigan, or eastwards into the Mohawk-Hudson River Valleys along the southern margin of the ice lobe in the Ontario

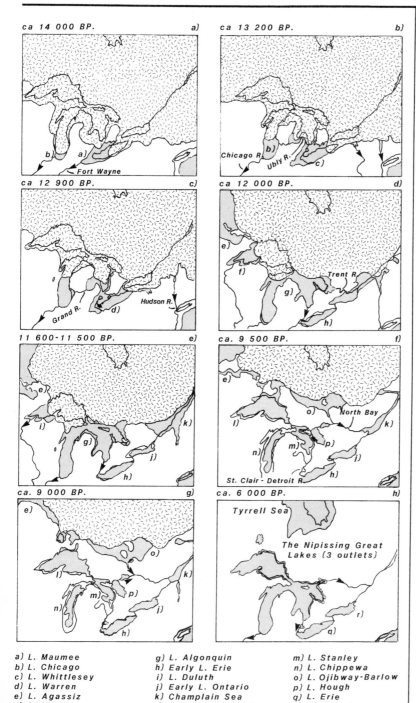

ca 14 000 BP.   a)
ca 13 200 BP.   b)
Chicago R.   b)   Ubly R.   c)
— Fort Wayne
ca 12 900 BP.   c)
0
Grand R.   d)   Hudson R.
ca 12 000 BP.   d)
e)   f)   Trent R.
g)   h)
11 600-11 500 BP.   e)
e)   i)   g)   0   k)   j)   h)
ca. 9 500 BP.   f)
e)   l)   o)   North Bay   k)   n)   m)   p)   j)   h)   St. Clair - Detroit R.
ca. 9 000 BP.   g)
e)   l)   o)   k)   n)   m)   p)   j)   h)
ca. 6 000 BP.   h)
Tyrrell Sea
The Nipissing Great Lakes (3 outlets)
r)   q)

Figure 5.9.
Stages in the development of the Great Lakes (Prest 1970). (BP: before present.)

a) L. Maumee
b) L. Chicago
c) L. Whittlesey
d) L. Warren
e) L. Agassiz
f) L. Keweenaw

g) L. Algonquin
h) Early L. Erie
i) L. Duluth
j) Early L. Ontario
k) Champlain Sea
l) L. Minong-Houghton

m) L. Stanley
n) L. Chippewa
o) L. Ojibway-Barlow
p) L. Hough
q) L. Erie
r) L. Ontario

ice   water   ► spillway

Basin. Ice retreat also allowed Lake Keweenaw to form in the western Superior Basin.

Lake Algonquin occupied most of the Michigan and Huron Basins between 12,500 and 12,400 years ago. Early Lake Algonquin (184 m asl) may have drained through the Chicago outlet at first and, somewhat later, into Early Lake Erie (142 m asl), Lake Iroquois (possibly 102 m asl) in the Ontario Basin, and the ice-free Mohawk-Hudson system. The level of Lake Algonquin fell to about 177 m asl (today's level) about 12,400 years ago, when ice retreat exposed a new outlet through the Trent River Valley, Kawartha Lakes, Kirkfield, and Fenelon Falls into Lake Iroquois. By about 12,000 years ago, however, lake levels were again at about 184 m asl, when Lake Algonquin was using the Chicago River and the Trent and St Clair drainage systems (Fig. 5.9d).

Ice re-occupied the Superior Basin and most of the Michigan Basin during the Valders advance, about 11,800 years ago. Lake Algonquin was largely restricted to the Huron Basin, discharging through the St Clair channel into Early Lake Erie, and thence into Early Lake Ontario and the Champlain Sea. As the ice retreated between about 11,600 and 11,500 years ago, Lake Algonquin expanded back into the Michigan Basin and Lake Duluth occupied the western Superior Basin (Fig. 5.9e). Further ice retreat opened up new glacially depressed outlets between 11,200 and 10,900 years ago. In the Michigan-Huron Basins, the post-Algonquin Lake drained to very low levels through the French River-Lake Nipissing-Mattawa River Valleys or the North Bay outlet to the Champlain Sea. By about 10,600 years ago, the deeper portions of the Michigan, Huron, and Georgian Bay Basins were occupied by Lakes Chippewa, Stanley, and Hough, respectively. These lakes drained through Early Lake Nipissing into the receding Champlain Sea. Rising water levels over the next few thousand years eventually caused Lakes Nipissing, Stanley, and Hough to merge, forming a single lake centred on the Huron Basin (Fig. 5.9f,g). Lake Minong, and then at a lower level Lake Houghton, occupied almost all the Superior Basin between about 10,300 and 7,800 years ago.

Continued isostatic uplift of the northern parts of the basins caused an expansion of the upper lakes, and the formation, between about 6,000 and 5,500 years ago, of the Nipissing Great Lakes (184 m asl) (Fig. 5.9h). Because of the uplift of the North Bay spillway, the Chicago River and St Clair outlets were also being used at this time, and eventually only the two southern outlets could be employed. Downcutting of the St Clair outlet and rock control of the Chicago outlet led to the adoption of a single outlet about 5,000 years ago, lowering Lakes Huron and Michigan to their present elevation (176.78 m asl) (Prest 1970, Karrow and Calkin 1985).

Lake Agassiz

Lake Agassiz was the largest of all the glacial lakes in North America (Fig. 5.10). It extended over a total area of almost 950,000 km² in North Dakota,

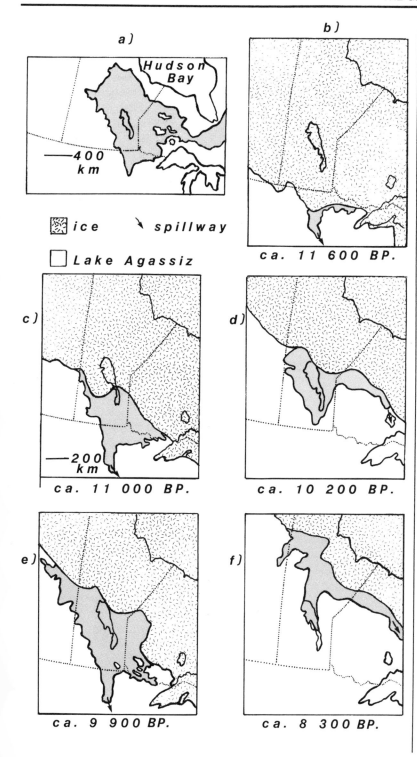

a)

Hudson Bay

—400 km

ice ↘ spillway

Lake Agassiz

b)

ca. 11 600 BP.

c)

—200 km

ca. 11 000 BP.

d)

ca. 10 200 BP.

e)

ca. 9 900 BP.

f)

ca. 8 300 BP.

Figure 5.10.
Stages in the development of Lake Agassiz (from Teller, in Karrow and Calkin 1985).

Minnesota, Saskatchewan, Manitoba, and Ontario, although not all this region was covered at any one time (Teller and Clayton 1983). Lake Agassiz began to form about 11,700 years ago, as meltwater from the retreating Laurentide ice sheet was impounded between the ice margins and the Manitoba Escarpment. The position of the ice determined the outlet channels for the lake, and a general fall in lake levels has been attributed to the successive use of spillways to the south, northwest, east, and north. The lake disappeared between 7,800 and 8,000 years ago, when the disintegration of the ice sheet allowed it to drain into Hudson Bay. Lakes Manitoba, Winnipeg, Dauphin, and Winnipegosis are remnants of the much more extensive lake that once existed in this region.

Lake Barlow-Ojibway and Other Large Lakes

Three to four thousand years (about 10,600 years ago) after the retreating Laurentide ice sheet had exposed the southern basins of the Great Lakes, meltwater was being impounded in northern Ontario between the ice front and the higher land to the south (Fig. 5.9). Lake Barlow discharged into the Ottawa Valley through the deep trench that today contains Lake Timiskaming. Lake Ojibway developed further west, but it soon joined up with Lake Barlow to form a single lake that was up to 960 km in length. Outlets were used at various times through the Timiskaming Valley, into the Superior basin, and northwards into Hudson Bay. Lake Barlow-Ojibway disappeared about 7,900 years ago, when disintegration of the ice allowed it to drain into Hudson Bay.

Another large glacial lake was created by meltwater flowing from retreating Keewatin ice, when its margins were along the edge of the Shield. Glacial Lake McConnell consisted of a series of interconnected lakes extending from the Great Bear Lake Basin, through the Great Slave Lake Basin, to the lower Athabasca and Peace River Valleys. Great Bear Lake, Great Slave Lake, and Lake Athabasca today occupy the largest closed basins of former Lake McConnell. These basins were excavated by glacial erosion of the weaker Phanerozoic rocks at the edge of the Shield. Other large glacial lakes developed on the Shield west of Hudson Bay, particularly in the basin of the Thelon River, where the ice blocked the eastward-flowing drainage routes, and in northern Québec-Ungava.

Glacial Sediments and Landforms

6

A very large proportion of Canada's landforms and **sediments** are primarily glacial in origin, although they may have been modified by other mechanisms. Many other features that do not have glacial origins have been modified by glacial processes. There are glacial **erosional** and **depositional** features in virtually all areas of the country, but erosion was generally dominant in the central portions, whereas most deposition took place around the outer margins.

Till

Till is poorly sorted, glacially deposited sediment (Chapter 4). There are many kinds of till, but they often consist of sub-angular rock fragments or clasts, ranging in size up to boulders, within a finer matrix of sand, silt, or clay. The kind of till in an area is determined by the type and structure of the bedrock, the way the **debris** was carried by the ice, and the **unconsolidated** sediments that are incorporated into it. Although the character of most tills is intimately related to the underlying bedrock, they can contain a variety of rock types and minerals derived from a wide area. The characteristics of a till change with the distance of travel, the rock fragments being progressively worn down into small mineral grains as they are carried by the ice. In some tills, the tendency for the long axes of rock fragments to be oriented parallel or perpendicular to the direction of ice movement (a till fabric), can provide valuable clues to the glacial history of an area.

Glacial **erratics** are rock fragments carried by ice from their place of origin and left in an area where there is a different type of bedrock. Erratics in tills are invaluable in determining the direction of former ice movement. Our present views on the structure and flow patterns of the Laurentide ice sheet in the late **Wisconsin**, for example, owe a great deal to the evidence accrued from the distribution of erratics (Chapter 5). Some of the most remarkable erratics are in Alberta, in a narrow zone extending southeastwards from the Jasper area to beyond the international boundary. The Foothills Erratics Train contains tens of thousands of pinkish or purplish quartzite boulders, one of which weighs more than 16,000 tons. These erratics are thought to have been carried by the fast-flowing ice stream produced by the confluence of Laurentide and Cordilleran ice (Dyke and Prest 1987). Erratics may be distributed in long, narrow dispersal trains extending back to their source areas. These trains may have been the result of fairly fast-moving streams of ice within a generally more slowly moving ice sheet, or the occurrence of normal but sustained flow over a fairly small source area (Dyke and Morris 1988).

Most of the tills in Canada were laid down in the Wisconsin glacial stage. Older tills lie beneath these deposits in some areas, and they are exposed in the Arctic Islands and the Yukon, beyond the limits of Wisconsin ice. Seven till provinces have been distinguished in Canada (Fig. 6.1) (Scott 1976):

(a) The Appalachian Province has two subregions: the lowlands of the Eastern Townships of Québec and southwestern New Brunswick, which were covered by Laurentide ice; and the uplands of Newfoundland, Nova Scotia, Gaspé, and New Brunswick, which were affected by local ice caps as well as, at times, Laurentide ice. In the Québec region, lodgement tills deposited at the base of Laurentide ice consist of stones in a matrix of approximately equal amounts of sand, silt, and clay. This is overlain by

Figure 6.1.
The till provinces of Canada (Scott 1976).

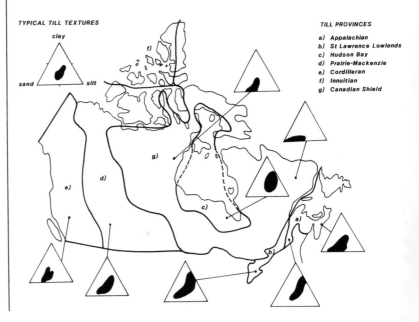

TYPICAL TILL TEXTURES

clay

sand   silt

TILL PROVINCES

a) Appalachian
b) St Lawrence Lowlands
c) Hudson Bay
d) Prairie-Mackenzie
e) Cordilleran
f) Innuitian
g) Canadian Shield

englacial till largely composed of boulders, although it is replaced by a thicker, sandy and bouldery **ablation till** in the mountains near the international border. Rapid changes in till types in the Maritimes reflect the variety of rock types, **intrusions**, and narrow rock outcrops in the Appalachians.

(b) The clayey tills of the St Lawrence Lowland Province are underlain in most places by Palaeozoic limestones and shales. The till cover over large areas of the eastern and central portions of this region was removed by the waves of the Champlain Sea, or overlain by marine clays. In southern Ontario, competing ice lobes occupying the bedrock depressions fluctuated back and forth, stacking successive till sheets on top of each other. Less than 15 per cent of the till content consists of metamorphic and igneous materials, although the proportion becomes much higher close to the boundary of the Shield. Lacustrine clay was incorporated into the tills near the lakes, and sandier **glaciofluvial** materials in the higher, more central regions.

(c) Tills in the Hudson Bay Province are similar to those in the St Lawrence Lowlands. Most of the material in the southern parts of this province appears to be lodgement till, although boulders on the till surface may have been deposited englacially. The tills in the northern part of the province were reworked by the waves of the Tyrrell Sea.

(d) The thick accumulations of till in the Prairie-Mackenzie Province are commonly composed of several sheets. Although ice flowed over this region from the Cordillera and Shield and from other directions, more than 80 per cent of the till is derived from local bedrock. These tills contain roughly equal proportions of sand, silt, and clay.

(e) It is difficult to categorize the tills of the Cordilleran Province because of complex patterns of ice build-up, and great changes in rock type and structure within short distances. While the range of till types is enormous, Cordilleran tills are generally compact and pebbly, with a sandy, silty matrix. In areas of low relief, the tills are usually 1-5 m in depth, but they can be thicker in valleys and depressions, and may be completely absent on bedrock knolls. In the mountainous areas of high relief, till can be up to 30 m thick in the bottoms and on the lower slopes of the valleys, but it is patchy or absent at higher elevations. Lodgement till is most common in the Cordilleran Province, although there are also shallow englacial deposits and boulder ridges.

(f) There has been little study of the tills of the Innuitian Province. The former glaciers of the High Arctic were probably cold and therefore quite passive, apart from those on Ellesmere Island. The till in most of this region is thin and discontinuous, consisting of a mixture of stones, fine sand, silt, and clay.

(g) Tills on the Shield tend to be coarse-grained and non-calcareous. They are usually between 2 and 8 m in depth, but can be thicker over the bedrock valleys. There are stoney lodgement tills in parts of the western

Shield and Labrador, and ablation tills, consisting of boulders in a sandy matrix, in the southern and eastern parts of the Shield.

Moraines

Moraines are ridges or mounds of glacial material deposited at, or close to, the ice margins. The term is used to encompass a vast range of features with different morphology and composition, and formed by different processes in a variety of depositional environments (Prest 1968).

Terminal or end moraines are composed of material that accumulated across the termini of actively moving ice. They can be formed in front of an advancing glacier or one which has been stationary for some time. Most major terminal moraines are therefore close to the former perimeter of the Laurentide ice sheet, and they are not common in the central zones where the ice retreated and finally melted.

Terminal moraines can consist of debris that was carried mainly at the bottom or on the top of the ice. Another type, however, was formed where frozen sediments were pushed and thrust upwards into faulted blocks by the advancing ice sheet. These ice-thrust ridges are quite common in the western Interior Plains and the Arctic; examples include Herschel Island and adjacent areas on the Yukon coast, and the Neutral Hills in Alberta. It is also likely that the Missouri Coteau escarpment in Saskatchewan was pushed up higher by the advancing ice (Fig. 1.8). Ice-thrust ridges and related features in central Alberta developed in topographic troughs where the permafrost in front of the ice sheet was thawed by **proglacial** water bodies, thereby reducing the resistance of the material to thrusting (Tsui et al. 1989).

Several features that could be mistaken for terminal moraines were not formed at the ice terminus. Some moraine-like forms in the western Interior Plains, for instance, are essentially ridges of bedrock hidden beneath a cover of till. Other ridges were formed where slabs of bedrock or frozen sediment were stacked on top of each other as a result of **shearing** within the ice.

Kame and Delta Moraines

Kame moraines consist of sediment deposited at the ice margins by meltwater streams, rather than directly by ice. Delta moraines (or flat-topped moraines) are also formed by meltwater streams, but along ice fronts that were standing in water. The western part of the St Narcisse Moraine in Québec and the moraines running along the northern side of the Canadian Appalachians are, in part, delta moraines (Fig. 6.2). Another large delta moraine was deposited in Lake Barlow-Ojibway, north of Cochrane and northeast of Kapuskasing.

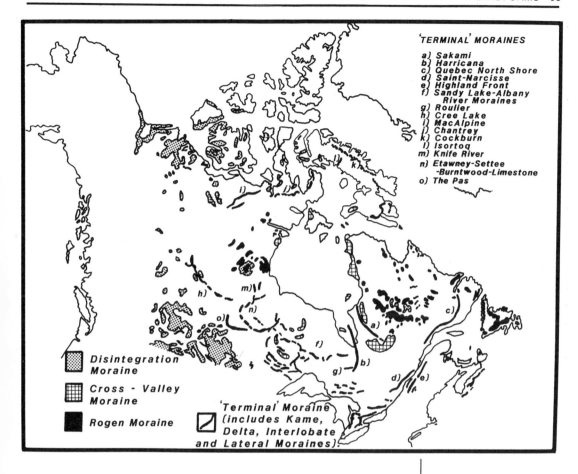

'TERMINAL' MORAINES

a) Sakami
b) Harricana
c) Quebec North Shore
d) Saint-Narcisse
e) Highland Front
f) Sandy Lake-Albany
    River Moraines
g) Roulier
h) Cree Lake
i) MacAlpine
j) Chantrey
k) Cockburn
l) Isortoq
m) Knife River
n) Etawney-Settee
    -Burntwood-Limestone
o) The Pas

Disintegration Moraine

Cross - Valley Moraine

Rogen Moraine

'Terminal' Moraine
(includes Kame, Delta, Interlobate and Lateral Moraines)

Figure 6.2.
Moraines in Canada.

Re-equilibrium Moraines

Terminal moraines provide a valuable record of advances and major **still-stands** in the position of the ice margin. While they may therefore reflect changes in climate, this assumption is not always justified. Many moraines in Canada are found just within the area that was covered by proglacial lakes or the sea, or just to the landward side. Ice margins in contact with water may have suddenly become grounded on land, as a result of glacial retreat, a break in the slope of the ground, or a drop in the water level. Until the ice was able to adjust to the reduced rate of ablation, its rapid flow, which was no longer required to compensate for rapid **calving**, could cause it to advance or become stationary. Moraines formed at the ice terminus under these conditions have been termed 'moraines' (Hillaire-Marcel and Occhietti 1980, Occhietti 1983). It has been suggested, for example, that the Sakami Moraine, running for 600 km to the east of James Bay, was formed when Lake Ojibway drained into the

Tyrrell Sea (Fig. 6.3) (Hillaire-Marcel et al. 1981). There is a continuing debate on the possibly similar origin of a number of other moraines, including the Isortoq and parts of the Cockburn Moraines on Baffin Island, the Roulier Moraine on the northern edge of Lake Barlow, and portions of the St Narcisse Moraine and the Québec North Shore Moraine System in southeastern Québec (Fig. 6.2).

Figure 6.3.
The re-equilibrium origin of the Sakami Moraine, Québec, following the drainage of Lake Ojibway into the Tyrrell Sea. The moraine was formed as a result of the stabilization of the ice margin until the excessive ice between profiles a and b had disappeared (Hillaire-Marcel et al. 1981).

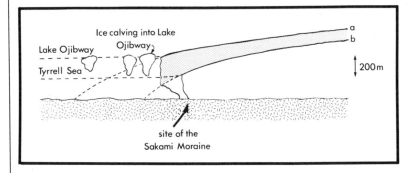

Interlobate and Radial Moraines

The underlying topography had a strong influence on the form of the ice margin as it retreated and thinned towards the close of the last glacial stage. The ice assumed a lobate shape in many parts of the country, as tongues of ice occupied the topographic depressions while the intervening areas of higher ground became ice-free. Interlobate moraines were formed by the deposition of sediment between adjacent ice lobes, often by running water. Many interlobate moraines are therefore also kame moraines. Examples include the esker-like Harricana Moraine, which developed between Labrador and Hudson ice, and the Knife River Moraine and the Etawney-Burntwood morainal complex of northern Manitoba, which developed between the Hudson and Keewatin ice masses (Chapter 5) (Fig. 6.2).

In northern Manitoba, kame moraines oriented at right angles to the former Keewatin ice terminus are thought to have developed at the edge of fast-moving ice streams within otherwise sluggish or thin ice. These radial moraines were deposited by meltwater flowing within the ice or in open channels, in part as deltas in Lake Agassiz (Dredge et al. 1986). A long till ridge on Victoria Island is also thought to be a type of lateral moraine formed between a fast-moving ice stream and cold-based ice with no **basal** movement.

Moraines in Canada

Large terminal moraines are uncommon in most of the Canadian Arctic, with the exception of Baffin Island and the northwestern perimeter of the Laurentide ice sheet. The term 'Cockburn Moraine System' has been proposed for a major series of moraines that run for more than 2,000 km along the northeastern coast of Baffin Island, down the western coast of the Melville Peninsula, and across northern Keewatin (the Chantrey and MacAlpine Moraines) (Falconer et al. 1965). This system, which appears to mark the position of a major stillstand or ice re-advance between about 8,600 and 8,400 years ago, may be correlated with the Cree Lake Moraines and the Sandy Lake-Albany River Moraines of northwestern Ontario (Fig. 6.2). Other moraines on Baffin Island represent the former margins of the Penny, Barnes, and Hall icecaps (Bird 1967).

The term 'Horseshoe Moraines' refers to a series of moraines built in the form of a horseshoe around the higher ground in the centre of southwestern Ontario, the site of the former Ontario Island (Chapter 5) (Fig. 6.4). The Orangeville Moraine consists of glaciofluvial gravel and **glaciolacustrine** sand deposited in an interlobate position between the re-advancing Georgian Bay and Ontario ice lobes about 15,000 years ago. The sand and gravel in the large Oak Ridges Moraine was formed a couple of thousand years later, as the outwash of a braided stream system, north of the advancing Ontario ice lobe (Gwyn and Cowan 1978). Part of the deposit was overridden and covered in till by the advancing ice. The Dummer

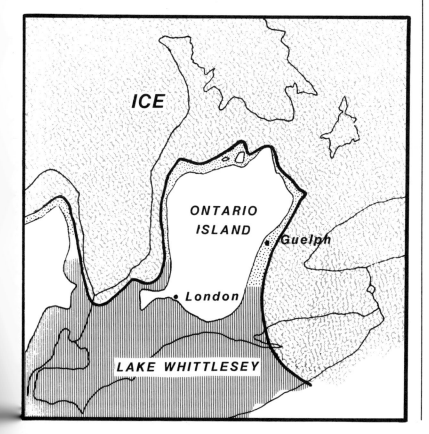

Figure 6.4.
The deglaciated Ontario Island (Chapman and Putnam 1966).

Figure 6.5.
The moraines of southern Ontario (Chapman and Putnam 1966).

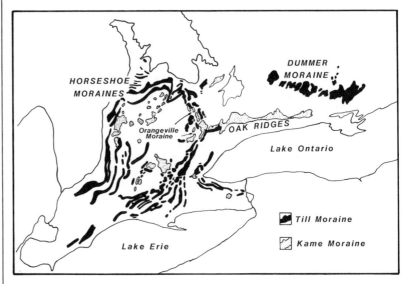

Moraines along the edge of the Shield are the youngest moraines in southern Ontario (Fig. 6.5).

The retreating Laurentide ice sheet left a series of morainal complexes in southern Québec, many of them deposited by, or in, water (Fig. 6.2) (Gadd 1964, Dubois and Dionne 1985). The term 'Highland Front Moraine' has been used for a system extending for about 362 km between Granby and Rivière du Loup. It has been suggested, however, that the term should be abandoned, as it applies to a variety of glacial, glaciofluvial, and **glaciomarine** deposits and landforms, which were formed at different times (Parent and Occhietti 1988). There are several smaller, and presumably older, systems on the higher ground to the south, including the partly deltaic Frontier Moraine. The Drummondville Moraine was built to the north of the 'Highland Front' during a halt in the recession of the ice. The ice then retreated to the northern side of the St Lawrence, where it built the St Narcisse Moraine during a re-advance about 11,000 years ago. The moraine is about 500 km long, extending from near the Saguenay River northeast of Québec to near Ottawa. Although it has been suggested that the St Narcisse is a re-equilibrium moraine resulting from calving in the Champlain Sea, this interpretation has been questioned, as only a small part of it was in contact with the sea. The 800 km long Québec North Shore Moraine, probably formed between about 9,700 and 9,400 years ago, is also a subject of dispute as workers disagree on whether a small part of it can be explained by the re-equilibrium hypothesis, although some sections were deposited in the Goldthwait Sea.

Small or Minor Moraines

There are several types of small morainal ridges that are parallel to the former ice terminus, but two appear to be particularly common:

Rogen (ribbed, rib, ripple) moraines consist of a series of short, sinuous ridges, approximately normal to the direction of ice flow. They may be asymmetrical, with gentler **stoss**, or up-ice, slopes than lee slopes. The ridges are usually less than 10 m in height and 2 km in length, and they are often separated by lakes or a **felsenmeer** of shattered rock (Chapter 7). Ridge crests can be fluted or drumlinized, and bands of Rogen moraine commonly alternate with drumlinoid topography along a line perpendicular to ice movement.

Rogen moraines are generally most extensive in the central portions of glaciated regions. They have been reported from Newfoundland, northern Québec and Labrador, northern Manitoba, southern Keewatin and the middle Thelon-Back River divide, and in a few places on eastern Victoria Island (Fig. 6.2). They tend to be found in shallow depressions or at the bottom of broad valleys, although in some places around the Keewatin ice divide they occur on the top of hills. The association of Rogen moraines with drumlins and other streamlined features suggests that they are not formed by stagnant ice. Some workers believe they develop where shear planes leave the floor in ice with a large amount of debris in the basal layers. This could be the result of normal compressive ice flow (Chapter 4), or the effect of bedrock obstructions or old river valleys (Fig. 6.6).

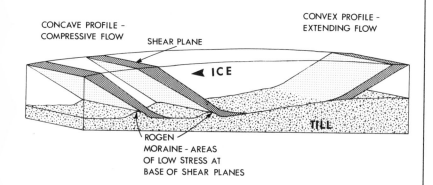

Figure 6.6.
A theory for the origin of Rogen moraine.

Cross-valley moraines (which are similar to De Geer Moraines) are narrower and have a more delicate appearance than Rogen moraines. They are regularly spaced ridges up to 300 m apart, but usually less than 15 m in height. They are thought to have developed at the front of, or beneath, a retreating ice margin that was standing in a lake or the sea. Some workers think they developed annually, as the ice front retreated. The best examples are on the eastern side of Hudson Bay, but they are also found in many other areas, including the western side of Hudson Bay, northeastern New Brunswick, the area northwest of Lake Nipigon, the Arctic mainland around Chantrey Inlet, the Boothia Peninsula, northwestern Southampton Island, and Prince of Wales Island (Fig. 6.2).

Ice Stagnation (Dead-Ice Wasting)

A glacier moves as it transfers excess snow and ice from the accumulation zone above the snowline to the ablation or melting zone below. Movement ceases in an ice mass that no longer has an accumulation zone. Reasons for this include the snowline's rising above the ice surface at the end of a glacial period; the isolation and separation of a portion of a retreating glacier on high ground; or the general thinning and disintegration of the ice into many smaller blocks. However, ice can also become stagnant, or virtually stagnant, in the marginal zone of an active, retreating glacier.

Deglaciation over large areas, particularly in western Canada, was accomplished by ice-thinning and disintegration, rather than by the steady retreat of the ice margins. Disintegration features, deposited during the last phases of glaciation, formed over broad areas in the marginal zones of the ice sheet, and over uplands where the ice was thin. They are found in areas of high and low relief, and are often superimposed upon older, active ice forms, including drumlins and terminal moraines. The type of landform that developed depended upon such factors as the subglacial topography, the amount of meltwater and debris, and where the debris was carried in the ice. This section is concerned with ice stagnation features largely consisting of till, and especially with their form and occurrence in western Canada. Other stagnation features, such as kames and eskers, which consist of **stratified** ice-contact sediments formed where meltwater was abundant, are discussed later.

A confusing array of terms has been used to describe the forms of ice disintegration in western Canada. In many cases, however, the dominant features are a variety of ridges and plateaux. Two main landform categories have been distinguished. Uncontrolled forms are produced when forces are equal in all directions, resulting in round, oval, hexagonal, or polygonal features with a lack of linear elements. Linear or curved controlled forms are deposited in open **crevasses**, thrust planes, or other lines of weakness in the ice that were inherited from the time when it was still active. Controlled forms therefore tend to be oriented parallel, perpendicular, or at a 45° angle to the direction of former ice flow.

Uncontrolled knob and **kettle** or hummocky disintegration moraines consist of numerous small hills, mounds, and irregularly-shaped depressions. The plan shape is more rounded than in terminal moraines formed by active ice, and lacks the latter's distinctive linear trends. Knob and kettle moraine is found over large areas of western Canada, including Turtle Mountain in Manitoba, and the Missouri Coteau and Moose Mountain in Saskatchewan (Fig. 1.8). Moraine plateaux are fairly flat-topped or saucer-shaped hills within knob and kettle moraines, rising as high as the till knobs or even slightly higher. They consist largely of till, but there may be a thin layer of lacustrine silt or clay on the surface, enclosed within a ridge running along the outer rim. Plains plateaux in areas of gentler relief are smaller but more numerous than moraine plateaux. Ridges in the shape of irregular rings are common in the western Interior Plains in hummocky moraine and in areas of lower relief. These closed disintegration ridges can enclose depressions (rimmed kettles), mounds of glacial material, or moraine plateaux. The Prairie mound is a similar doughnut-shaped feature, about 100 m in diameter and several metres in height. Unlike that of rimmed kettles, however, its central depression does not extend below the general ground level.

A variety of controlled disintegration ridges in the western Interior Plains were superimposed on sediments originally deposited by active or stagnant ice. Straight or slightly arcuate linear ridges, up to about 10 m in height, 8 to 100 m in width, and from a few metres to many kilometres in length, are thought to have resulted from the filling of ice crevasses from above. Two sets of ridges generally intersect to form a box, diamond, or waffle pattern. They are found in areas of high relief, but are generally more regular where the terrain is fairly subdued.

Spectacular meltwater channels were cut in the western Interior Plains by streams flowing in tunnels or ice-walled trenches within stagnant ice. They are often marked by chains of depressions or kettles in areas of high relief, where the channels were floored by till, and broad, open troughs in areas of low relief, where the till cover is thin.

In some areas most of the sediment in disintegration features was laid down from above as the ice melted, while in others it was squeezed up into openings in the base of the ice. Ring and other ridge-like features, for example, could be formed by material sliding off melting blocks of stagnant ice; this appears to have been the origin of ablation slide moraines in the Northwest Territories. On the other hand, they could be produced by material squeezed up into subglacial tunnels, or into the gaps between blocks of ice. This has produced features in south and central Alberta that are very similar to glaciofluvial eskers, except that they consist of, or have a core of, till. No single explanation, therefore, can account for the variety of landforms and sediments produced by ice disintegration. Although material generally appears to have been laid down from on top of the melting ice in south central Saskatchewan, for instance, rings and ridges have formed in sub-, en- and supraglacial till, as well as in glaciolacustrine and glaciofluvial environments (Parizek 1969).

Streamlined and Ice-Moulded Features

Groups or fields of grooves and ridges of various types give a fluted appearance to many glaciated areas. These ice-moulded forms can be produced by erosion or deposition, or by a combination of the two, although they all tend to have a streamlined shape, elongated in the direction of ice flow (Fig. 6.7).

Erosional Features in Rock

Large areas of the Canadian Shield consist of irregular rock basins and lakes, scoured out by ice along joints, faults, and other rock weaknesses.

Figure 6.7.
Glacially streamlined features.

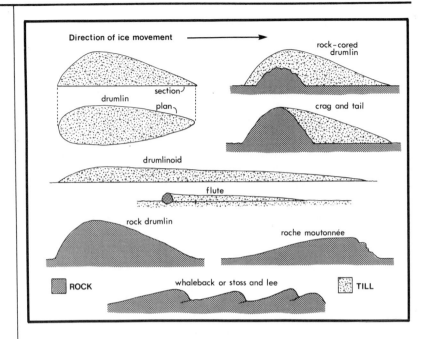

The intervening areas of higher ground may be in the form of whalebacks (stoss and lee forms), rock drumlins, or roches moutonnées (Dionne 1987). Mathematical modelling suggests that these features are genetically related, reflecting different levels of **stress** exerted against bedrock obstructions by moving ice (Fig. 4.4). The particular shape that develops – especially whether it is symmetrical, or steeper on the upstream or downstream side –may therefore depend upon such factors as ice thickness and velocity, and the shape of the original rock obstruction (Boulton 1974).

Moving ice armed with rock abrasives scratches and polishes the underlying bedrock. Striations or striae are small scratches, usually less than a metre in length, oriented parallel to the direction of ice movement. Grooves are essentially very large striae cut by boulders rather than small rock particles. Grooves in bedrock in Alberta, Saskatchewan, and the Northwest Territories are largely independent of rock structure and topography. Some extremely large grooves and residual ridges in bedrock west of Great Bear Lake are up to 90 m in width, 30 m in depth, and 13 km in length, although the average dimensions are much less. Striations are most prominent on fine-grained rocks, but a variety of crescentic-shaped gouges and fractures known as friction marks are usually more common on medium-grained rocks. Friction marks, which normally have long axes perpendicular to ice movement, are thought to have developed where very high stresses at the ice base were maintained for short periods beneath rock fragments. A variety of scallops, bowls, curved and sinuous channels, and sickle-shaped troughs have been cut into rock along the eastern edge of the Canadian

Shield (Bernard 1971, Shaw 1988), and although reports are few, they are probably quite extensive elsewhere in Canada (Shaw and Kvill 1984). The origin of these features, collectively known as p-forms, has not been determined, although they appear to be the product of erosion by a flowing medium at the base of the ice. Some workers have attributed them to abrasion by ice and debris, while others believe they were produced by subglacial streams, through either cavitation or abrasion (Chapter 8).

Flutes, Drumlinoids, and Crag and Tails

The term 'glacial fluting' is commonly used to describe streamlined landforms, although it lacks precise definition. Some restrict its use to shallow and narrow grooves, usually accompanied by adjacent ridges, while others have included low drumlinoids and giant grooves. Small flutes, ridges up to a few metres in height and width but sometimes up to 1 km in length, are formed when wet till is squeezed up into cavities engraved in the base of ice as it passes over bedrock projections or boulders (Fig. 6.7). It has been shown that the random occurrence of boulders can account for the apparently fairly regular spacing between flutes in many areas.

Much larger ridges have also been called flutes by some workers, and drumlinoids by others. Large flutes can be about 25 m in height, 100 m in width, and 20 km in length. The ridges in Alberta and central Québec-Labrador tend to be about 90 to 120 m apart. It is questionable whether till could be squeezed upwards under sufficient pressure to form these large streamlined ridges, or whether cavities could extend far enough up into thick, flowing ice. Explanations for the origin of these features are similar to those proposed for drumlins, and no less controversial. It has been suggested, for example, that they are the result of zones of high and low pressure at the ice base, or secondary flow within the ice in the form of alternating cells; these cells could have been generated where the flow of the ice converged in the low pressure zones down-glacier of ice-thrust blocks or other topographic obstructions (Moran et al. 1980, Jones 1982).

The term 'drumlinoid' has been used to describe very long and narrow drumlin-like ridges, a few kilometres in length and from about 3 to 30 m in height. They can be tapered at both ends, or the stoss slope can be blunter than the lee slope. Some have bedrock knobs at their upstream ends. They are found in most glaciated regions, although they are often difficult to see on the ground. There are enormous fields on the Shield in northern Québec and between Hudson Bay and Great Slave and Athabasca Lakes, particularly within and down-glacier of the Thelon and Athabasca Plains, where the Proterozoic basins provided large amounts of sediment (Figs. 1.4 and 1.7).

Crag and tail formations can be considered a special type of drumlin, in which a tail of sediment extends downstream from a projecting rock knob (Fig. 6.7). There are many normal crag and tails on Victoria Island, but a related form on the northern coast consists entirely of rock, the tail being composed of rock that is less resistant than that in the crag. Crag and tails are usually found in areas where the **drift** is thin. In the central Lake Plateau region of Labrador-New Québec, for example, there are drumlins and drumlinoids on the drift-covered plains and in the major valleys, and crag and tails on the uplands.

Drumlins

The classical drumlin has the half-ellipsoid shape of an inverted spoon, with a long axis parallel to the direction of ice movement (Fig. 6.7). Drumlins can occur as single hills, but are usually found in groups or fields containing large numbers of individuals. Drumlins and drumlinoids can be composed entirely of till, but many contain at least some stratified, water-laid material.

Most drumlins are probably between about 15 and 40 m in height, but some reach more than 70 m. They are higher and more oval in shape than flutes or drumlinoids, with typical length-to-width ratios of between 1:2 and 1:4; this compares with a corresponding ratio for drumlinoids of between 1:15 and 1:30. The stoss side of classical drumlins is wider and steeper than the lee side, although the slopes are approximately equal on many drumlins, and in a few cases the lee slope is steeper. Few drumlins possess all the characteristics of the classical form, and there is usually considerable variation in their shape and size within single fields. Some variations in the Guelph and Peterborough drumlin fields of Ontario seem to be random, but others appear more systematic, and may be related to the progressive slowing and thinning of the ice as it travelled towards the glacial terminus (Trenhaile 1975, Crozier 1975).

There are large numbers of drumlins in the Lunenburg area and in the smaller fields of western Nova Scotia; in southern Ontario between Lake Simcoe and Trenton, and in several other fields; in a number of small fields in Manitoba, Saskatchewan, and Alberta; and on the Interior Plateau of British Columbia. There are also drumlin fields on Victoria, Prince of Wales, and King William Islands, as well as in a few areas on the north coast of the District of Keewatin. Most drumlins, therefore, were formed fairly close to the margins of the Laurentide ice sheet at its maximum. Although there are individual drumlins and groups of drumlins in central Labrador, in northern Ontario, and in other central areas covered by the Laurentide ice sheet, drumlinoids are far more common.

Differences in drumlin morphology and composition probably reflect the fact that different processes and combinations of processes can produce similarly shaped landforms. Many theories have been advanced to account for the formation of drumlins, although none of them provides a totally adequate explanation (Menzies 1979, 1989):

(a) Some drumlins seem to have been built up in layers around a pre-existing drift or rock knoll, possibly even around obstacles created by frost heave.

(b) The dilatancy theory suggests that drumlins form as till attempts to dilate or expand beneath glacial ice; this occurs under certain critical ice-pressure conditions, producing stable mounds of subglacial material around which drumlins can develop. According to this theory, drumlins form where these critical pressures occur, in an intermediate zone in an ice sheet where the stress is too low to permit complete erosion of the basal till, yet high enough to streamline it. High stress under thick ice in the interior of an ice sheet results in erosion and transportation but little deposition, whereas low stress under the thin ice near the ice margins permits deposition but not streamlining. An alternative suggestion is that drumlins fail to develop in the central portions of an ice sheet not because the ice is too powerful, but because the strength of the basal till under high pressure is too great.

(c) Some drumlins and large flutes may be the result of secondary flows in glaciers. Drumlin fields in Alberta, for instance, occur where the topography would have caused ice flow to converge, possibly generating secondary flows.

(d) Drumlins may also be formed of sediment squeezed up or washed into cavities at the ice base. Examples include drumlins consisting of stratified material in northern Saskatchewan, which have been attributed to the infilling of cavities by meltwater (Shaw and Kvill 1984). It is also possible that some drumlins are the ridge-like remnants of glacial material that was eroded by subglacial meltwater (Shaw and Sharpe 1987).

Many drumlins have grooves and flutings superimposed on them, and it has been proposed that the often distinct progression from large flutes through drumlinoids to drumlins may reflect a decrease in the rate of ice movement. Other factors could also be important. It has been suggested that drumlinoids rather than drumlins formed in Keewatin and other areas where the till is rather thin, and either non-adhesive or very adhesive and clayey. The drumlins of central British Columbia are found in soft sedimentary or weak metamorphic rock areas, the crag and tails where there is more resistant granitic or metamorphic bedrock, and the grooves where the bedrock is flat. Well-developed drumlins and crag and tails are also found where the ice moved up a slope, but long, low drumlins and grooves are more common where it moved downslope. Drumlins and crag and tails also seem to be more prominent where the ice was thick than where it was thin.

Water-laid Deposits and Landforms

The amount of meltwater flowing from a glacier depends upon such factors as the ice temperature, the season of the year, and the permeability of the underlying bedrock. Water can exist in the liquid state only if the temperature of the ice is near to the melting point. No meltwater can exist at the bottom of cold ice, and even on the surface and at the ice margins, any meltwater activity is seasonal and local. Most of what follows in this section, therefore, refers to the effects of meltwater associated with warm ice. Large amounts of water flow in streams or sheets on the surface of warm glaciers, although it usually doesn't travel very far before disappearing into the ice through crevasses or cylindrical melt holes (moulins). Water flows in tunnels or conduits within the ice, and in films or in channels cut in the ice or rock at the base. Lakes may exist beneath the ice in deep depressions in the glacial bed. Meltwater flows away from the ice margins in streams or, if the terminus is standing in water, into ice-dammed lakes or the sea.

Meltwater streams, ice-dammed lakes, and glacially induced marine invasions have played a prominent part in sculpturing Canada's scenery (Figs. 5.8 and 6.8). Glaciofluvial, glaciolacustrine, and glaciomarine processes produced some unique landforms, but they have also often contributed to the development of moraines, streamlined landforms, and other features that are usually assumed to be deposited directly by glacial ice.

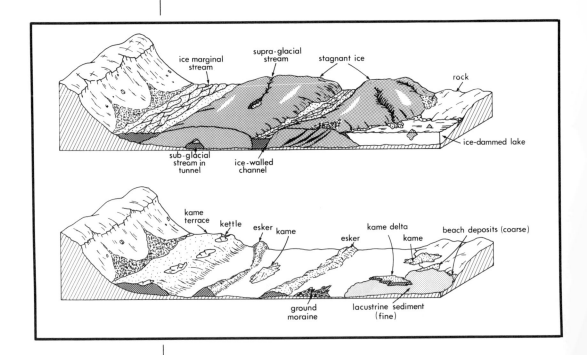

Figure 6.8.
The origin of some glacio-fluvial and glaciolacustrine features.

Meltwater streams can usually carry fine-grained silts and clays in suspension, depositing them in standing water or transporting them completely away from the region. Glaciofluvial sediment (stratified drift) therefore generally consists of beds of sand and gravel. The material is much better sorted than in glacially deposited tills, and the stones tend to be more rounded. The distinction is usually made between ice-contact sediments deposited on, in, under, or against the ice, and proglacial sediments deposited some distance from the ice margins.

Ice-contact Features

As well as sharing the normal characteristics of glaciofluvial sediments, ice-contact deposits show the effects of ice-melting, including twisted or contorted and faulted beds, and collapse structures.

Kames • The term 'kame' has been used to represent a wide range of ice-contact landforms, although most workers now use it only to describe isolated mounds of stratified ice-contact sediment. Kames consist of sand and gravel deposited by water under, in, on, or against stagnant, or almost stagnant, ice. It has been suggested that they form in ponds on the surface of stagnant ice, possibly as the deltaic deposits of meltwater streams. They could also be deposited directly onto the subglacial surface through a hole in the ice, or in cavities in or under the ice. Meltwater streams can also build kames at the ice margin, as fans growing out from the ice, or inwards against the ice. Large numbers of kames produce an irregular, hummocky landscape known as a kame complex or kame and kettle, consisting of a combination of kame mounds and kettle hollows.

Kame terraces • Ice-contact sediments were also deposited in stream channels or narrow lakes running between down-wasting ice-margins and the sides of valleys or hills. These sediments produced flat or irregularly topped terraces, sometimes with kettles where the supporting ice melted. Terraces can be distributed in a step-like sequence along valley sides if the ice down-wasted in stages. Kame terraces are widespread throughout Canada. In the interior valleys of British Columbia they were formed, along with fans and deltas, at the margins of stagnant ice lobes, about 50 to 100 metres above the valley floors (Figs. 8.4 and 8.21).

Eskers • Eskers are sinuous ridges of glaciofluvial material. They usually consist of gravel and cobbles, although sand is the main component in some instances. Swarms of eskers were formed where remnants of the Laurentide ice sheet wasted away, providing large amounts of meltwater. Eskers are found in southern and western Canada, but they are largest and most numerous in the north, in Labrador-New Québec, northwestern Québec, northern Ontario, eastern Victoria Island, and particularly in the western Canadian Shield, where they probably attain their greatest development.

Eskers are generally less than 50 m in height and 150 m in width, although some types are considerably wider. They range from a few hundred metres up to several hundred kilometres in length; the larger eskers usually have a few gaps where there was erosion or a lack of deposition. The Munro Esker in northern Ontario can be traced for 400 km, and the Thelon Esker extends for almost 800 km from Dubawnt Lake westwards to Great Bear Lake in the District of Mackenzie. The term 'crevasse filling' has been used to refer to shorter and straighter ridges of ice-contact sediment. These have been described in an area to the north of Great Bear Lake, where sand and gravel deposited in open fissures and crevasses form ridges resembling undulating to flat-topped eskers. A similar origin has been ascribed to a 'kame' ridge near Edmonton.

It has been suggested that eskers are deposited at the ice margin, or within about 3 to 4 km of it. This implies that very long eskers were constructed in stages as the ice margins retreated, and that their age must therefore increase downstream (Banerjee and McDonald 1975). Eskers tend to follow the course of valleys, although they can cross them if they meet at high angles. This may reflect the fact that subglacial streams were able to flow uphill for some distance, or it may be that the eskers were formed in, or on top of, the ice and were then superimposed on the underlying surface as the ice melted.

Eskers could have been formed in several ways:

(a) In stream channels contained within tunnels running through the ice or along its floor. Eskers that formed in englacial tunnels were later lowered onto the glacial bed as the ice melted.

(b) In open, ice-walled trenches on, or at the base of, ice sheets, or in ice-walled **re-entrants** in the ice margin. The Brampton Esker, in southern Ontario, was deposited in a glacial lake by braided streams flowing between the walls of a re-entrant in the Ontario ice lobe.

(c) As fans at the ice margin, or, if the terminus was standing in water, as deltas, which became elongated as the ice retreated.

It has been suggested that only stagnant ice can provide sufficient meltwater for the formation of eskers, and suitable conditions for their preservation. With this assumption in mind, it has been suggested that the ice was too active, while the Laurentide ice sheet still occupied areas well beyond the boundaries of the Shield, for meltwater channels to be maintained long enough for esker formation. By the time the shrinking ice sheet had become confined to the Shield, however, final deglaciation may have been characterized by down-wasting and ice stagnation, allowing esker deposition in tunnels at the ice base. This may explain why there are many more eskers on the Shield than in the surrounding areas. Nevertheless, different kinds of eskers required different conditions for their development, and there seems to be little reason why some types should not have developed in association with active ice.

There are three main types of esker:

(a) The embankment type is a single, continuous ridge with either a sharp crest or a flat top.

(b) Beaded eskers consist of a series of fairly regularly spaced, conical hills. This type of esker was deposited at a glacial terminus that stood in water, the beads being deposited annually during seasonal stillstands of the ice margin.

(c) Complex systems consist of broad ridges composed of many parallel ridges or crests. Very wide types were probably formed in ice-walled trenches open to the sky, in some cases as part of an interlobate morainal complex. The interlobate Harricana Moraine in the James Bay Lowlands, for example, is actually a very large esker (Fig. 6.2).

Ice temperature determines how much liquid exists within a glacier, and where it can flow within the ice mass. This may partly account for the occurrence of different types of esker throughout the country. In northern Manitoba, for instance, the eskers are long, continuous ridges running in roughly parallel swarms; they may have formed in extensive subglacial tunnels within warm ice. In southern Manitoba, the eskers occur as beads that are shorter, lower, and wider than those in the north. These eskers may have developed in braided streams flowing in open channels in the marginal zone of an otherwise cold-based glacier, where there was limited meltwater (Ringrose 1982).

Kame deltas • About 80 per cent of the retreating ice front of the Laurentide ice sheet terminated in lakes or the sea during the late Wisconsin. Streams flowing from the ice into these bodies of water produced an enormous number of deltas. These may contain ice-contact or proglacial sediments, depending upon whether the ice terminus stood in water, or whether the meltwater stream flowed on land for some distance before entering the lake or sea. A particularly large kame delta, extending for about 70 km from its apex near Brandon, Manitoba to near Portage La Prairie, was deposited in Lake Agassiz by the Assiniboine River. There are several large deltas in southern Ontario, including the Bothwell Sand Plain deposited in Lake Warren by the Thames River between London and Chatham, the Caradoc Sand Plains built by the Thames in Lake Whittlesey east of London, and the Norfolk Sand Plain built by the Grand River in Lakes Whittlesey and Warren south of Brantford. The many other delta deposits in southern Ontario include the Humber and Don River deltas in Toronto, which were built into Lake Iroquois, and the Petawawa, Barron, Indian, and Ottawa River deltas in the Pembroke area, which were built into the Champlain Sea during the Fossmill stage of Lake Algonquin (Chapter 5). Large glaciomarine deltas were built along the Bays of Chaleur and Fundy in New Brunswick, and in the Sept Iles area in Québec. Parts of the western St Narcisse Moraine in southern Québec consist of deltas built into the Champlain Sea by meltwater streams (Fig. 6.2). In northern Ontario, southeastern Québec, Baffin Island, and elsewhere, ice-contact deltas connected to eskers may reflect fluctuations, possibly seasonal, in the rate of retreat of the Laurentide ice sheet.

Proglacial Features

Enormous quantities of sand and gravel were deposited by braided proglacial streams flowing away from the ice terminus. The coarsest outwash material is normally deposited close to the ice and the finer sands further away. The simplest depositional form is that of a single fan of outwash, leading back to a supra-, en-, or subglacial stream. Most outwash, however, consists of a number of coalescent fans. Large fans extend through breaks in terminal moraines, and they can often be traced back into kame terraces and ground moraine. The distinction is usually made between outwash plains (plain sandur), where the deposits are able to spread out over a wide front, and outwash trains (valley sandur) confined within valleys (Church 1972). Valleys can be choked with sediment up to hundreds of kilometres from the ice front. In most cases, though, when the streams no longer have an adequate supply of glacial sediment, they cut into the outwash to form terraces (Chapter 8). Kettled or pitted outwash develops where stream deposition spreads back over buried or partially buried ice; kettle depressions then develop as the ice melts. If the kettles are very numerous, the destruction of the original outwash surface produces a landscape that is virtually indistinguishable, although genetically different, from kame and kettle. Most alpine valleys have train deposits, but they are also widespread in areas of lower relief, as in the valleys of the southern Canadian Shield between Ottawa and Québec, in the valleys on the Lake Plateau of Québec, and in part of the Annapolis Valley in Nova Scotia. Perhaps the largest outwash deposit in Canada is in the middle course of the Back River in the Northwest Territories (Fig. 8.2).

Alpine or Mountain Scenery

The most spectacular glacial scenery is usually found in rugged mountainous regions. The best examples in Canada are in the mountain ranges of the west (Ryder 1981, Luckman 1981), in eastern Baffin Island, and elsewhere in the highlands of the northeast. There are, however, smaller glaciated uplands and mountains in southeastern Canada, in the uplands of Gaspé, Cape Breton, and northern Newfoundland.

Cirques and Associated Features

Glacial accumulation zones in mountainous regions are often contained within deep hollows or cirques (cwms, corries), clustered around the sides and heads of valleys. Although the floors of many cirques slope outwards, the classical shape is generally considered to be a deep rock basin, backed by a steep headwall and with a residual lip or low rim of rock at the front. The lip is often buried beneath a terminal moraine (Fig. 6.9).

Cirques usually develop out of depressions that were initially excavated by streams, although they can form wherever a hollow is available to collect and retain snow. Cirques were progressively modified by glacial erosion, as they were successively occupied and abandoned throughout the Pleistocene. For example, on the basis of present rates of glacial erosion, it has been estimated that it would have taken between 2 and 14 million years to erode the cirques of Baffin Island (Anderson 1978). The floors of most cirques are close to, or just below, the glacial snowline or equilibrium line. Cirques that have now been abandoned by ice can therefore provide a rough estimate of the elevation of the snowline during glacial periods. It is not clear, however, whether the occurrence of a step-like series of cirques

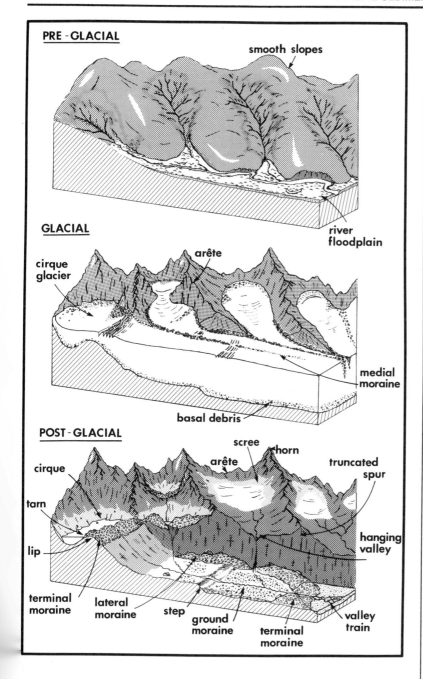

Figure 6.9.
The origin of some alpine glacial features.

differing in elevation by several hundred metres – in the Banff area, for instance – is the result of a number of glacial snowlines, or of the glacial exploitation of several favourable stream-cut hollows or geological sites (Fig. 6.10).

Figure 6.10.
A cirque staircase in the Rocky Mountains west of Banff, Alberta. The cirque containing Mummy Lake fed ice into the lower cirque containing Scarab Lake, which in turn fed the lowest cirque, which now contains Egypt Lake.

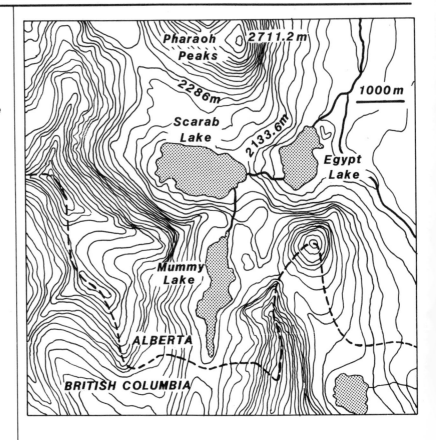

Probably the dominant erosive mechanism on the rock exposed above the ice surface is frost action. But if frost action lengthens and widens cirques, it is glacial erosion that lowers their floors. The basin shape of classical cirques is partly the result of rotational sliding of the ice. As the ice surface is steepened by the annual accumulation of snow and ice above the equilibrium line and by ablation below, the glacier must rotate to restore its equilibrium gradient. Abrasion beneath the rotating ice would therefore tend to excavate a basin, particularly as the greatest erosion probably occurs beneath the equilibrium line, where ice velocity and thickness are greatest.

Cirques in Canada and elsewhere in the mid-latitudes of the northern hemisphere tend to face between north and east. These hollows are shaded from the sun, which is in the south, and protected from the prevailing southwesterly winds, while they may gain large amounts of the snow that is blown from the mountain tops. Cirques in the southern Cordillera are therefore most numerous, best developed, and, because the snowline is lower, at lowest elevations on the north and northeastern flanks of the mountains (Trenhaile 1976).

The type of mountain scenery in a glaciated area is determined, to a large extent, by the density of the cirques. If they are far apart, the uplands retain much of their preglacial character. Alpine scenery is most dramatic where the original upland has been dissected by large numbers of closely spaced cirques. Where three or more cirques intercept, they form sharp-peaked mountains, or horns, with narrow ridges, or arêtes, radiating outwards from them along backwalls shared by pairs of cirques (Fig. 2.1). Mount Assiniboine, on the border between Alberta and British Columbia, is generally considered to be the best example of a glacial horn in Canada.

Valleys

Steep-sided glacial troughs were excavated by ice fed into valleys from cirques or mountain ice caps, and by fast-moving ice streams within ice sheets. Most troughs were preglacial river valleys that were inherited and modified by glacial action, although some were originally cut by glacial erosion. They are particularly common and spectacular in mountainous areas with heavy snowfall, where warm-based glaciers flowing down steep slopes accomplish large amounts of erosion. But troughs can also develop in lower areas. In southern Ontario, re-entrant valleys were cut by ice moving up and over the Niagara Escarpment. Although pre-existing cols and gaps were occupied and enlarged by the ice, some troughs may be entirely the result of glacial erosion. The Dundas Valley at the western end of Lake Ontario is a major glacial trough. In British Columbia, the Juan de Fuca and Georgia Straits were either formed or extensively modified by ice flowing off the western flanks of the Cordilleran ice sheet. Glacial scour is also thought to have been responsible for three major troughs on the continental shelf between the Queen Charlotte and Vancouver Islands.

The parabolic cross-sectional shape of many, though not all, glacial valleys is a flaring U-shape, whereas many valleys cut by streams are V-shaped. The floors of glacial valleys are generally flatter than river valleys, at least partly because of the deposition of glaciofluvial outwash train deposits. The floors of major valleys are deepened much faster by the large glaciers that occupy them than are the floors of tributary valleys with smaller glaciers. After glaciation has ended, the tributary valleys are therefore often left perched or hanging on the sides of the main valleys (Fig. 6.9). Picturesque waterfalls – such as the spectacular Takakkaw Falls in Yoho National Park, British Columbia – plunge from hanging valleys into the main valleys. In straightening their valleys, glaciers may cut off the ends of rock ridges to form truncated spurs (Fig. 6.9). The floors of glacial valleys often consist of alternations of glacially excavated basins and rock steps or bars (riegels) (Figs. 6.9 and 8.23). The elongated basins may contain lakes, particularly if the valleys are also dammed by moraines or other glacial debris. Examples in the Cordillera include Waterton Lakes, Maligne Lake, Lake Louise, and Peyto Lake in the Rocky Mountain system, though not, despite its name, Moraine Lake in Banff National Park, which was actually dammed by material from a landslide. Rock steps in the southern Canadian Cordillera sometimes occur where an increase in the thickness of the ice increased its ability to erode its bed. Some steps are therefore found at the junction of tributary valleys, or where the valleys bend or become narrower; many steps, however, are geologically induced. The occurrence of extending and compressive flow (Chapter 4) must also have played an important role in emphasizing basin and step relief.

Glaciated valleys contain a variety of glacial, glaciofluvial, and glacio-lacustrine landforms and sediments (Fig. 6.9). Kame terraces or lateral moraines develop between glaciers and valley walls. Lateral moraines are formed by scree and debris sliding and rolling off the ice surface. They range from thin veneers of debris on the valley side, up to well-defined ridges hundreds of metres in height. Lateral moraines usually increase in size where they merge into terminal moraines. Fresh lateral and terminal moraines in the southern Cordillera were formed during **neoglacial** ice advances. There have been several periods of ice advance in the last few thousand years (Slaymaker and McPherson 1977, Luckman 1981), but the most recent, and most extensive in many areas, may have culminated in the nineteenth century. Medial moraines often extend downstream on the ice surface from the point where two lateral moraines join at the spur of valley confluences. Essentially ice ridges protected from the sun's heat by a surface layer of debris, they are conspicuous features of contemporary valley glaciers, but most have not survived deglaciation in a recognizable form. Although streamlined depositional features are generally associated with continental glaciation, they are found in some alpine regions. They are common in the major valleys in the Jasper-Banff area, for example, extending into the Rocky Mountain Trench in the west and beyond the mountain front in the east. The low drumlinoids and flutes in this area are on the lower parts of the valley floors, with streamlined bedrock and crag and tails on the slopes.

Fiords

Fiords are drowned glacial troughs, with high, steep sides and irregular bottoms composed of shallow bars and deep rock basins. Rock bars at or close to the sea entrance are known as thresholds. There are numerous fiords along Canada's northeastern coast, from central Labrador to northern Ellesmere Island, and along about 850 km of the coast of British Columbia, where they are called channels, arms, passages, straits, canals, reaches, sounds, or inlets (Fig. 6.11). They average about 36 km in length on the Pacific mainland, with a maximum of about 130 km. Although there are about four times as many fiords in the Arctic as on the Pacific coast of North America, with the exception of those on Ellesmere and eastern Baffin Islands they tend to be shorter. Maximum depths range between 100 and 720 m on the mainland of British Columbia, and between 20 and 950 m on Baffin Island.

There are several kinds of fiord in the Canadian Arctic (Bird 1967):

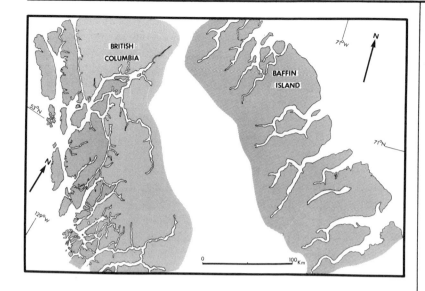

Figure 6.11.
Fiords in British Columbia and on the eastern coast of Baffin Island.

(a) Long Norwegian-type fiords have steep to vertical sides that are often more than 600 m in height.

(b) Fiords cut into horizontally bedded sedimentary rocks are wider and shallower than the Norwegian type. They are found on the southern and northwestern coasts of Devon Island, the southern coast of Ellesmere Island, and the western coast of Melville Island, and in southern Labrador around Hamilton Inlet.

(c) Cirque fiords originally contained cirque or small valley glaciers. They are only a few kilometres in length, and terminate abruptly landwards in vertical cliffs. They are found in northeastern Southampton Island, on the southern side of Hudson Strait, on the southern side of Frobisher Bay on Baffin Island, and in northern Labrador, as well as in the main fiord regions.

Although it is generally accepted that rivers and glaciers played an important role in excavating fiords, the coincidence of fiord patterns with geologic lines of weakness suggests that their work has been controlled by rock structure. Fiords radiate outwards from the Barnes Ice Cap on the eastern coast of Baffin Island, but the side valleys, which connect adjacent fiords, are parallel to the major northwesterly-trending faults. The fiords on the northern and southern coasts of Baffin Island are shorter, smaller in area, and straighter than those in the east. The orientation of the northern fiords, and to a lesser extent those in the south, also follows the north-westerly trend of the major faults (Bird 1967, Dowdeswell and Andrews 1985). It is not yet clear, however, whether these fiords are primarily the result of rivers and ice eroding joint and fault sets, or whether they are essentially long and narrow **grabens** or sunken areas between the faults.

The fiords of British Columbia also appear to be structurally controlled.

There are two main patterns. The dominant one consists of waterways running parallel and perpendicular to the coast. The subordinate pattern consists of linear sections oriented at an angle to the dominant pattern. These patterns seem to reflect joint orientations and other aspects of rock structure in this region. Although most of the fiords in western Canada probably began as structurally controlled river valleys in the late Tertiary, and were later subjected to extensive glacial excavation during the Pleistocene, some may have been cut entirely by glaciers.

How Deep Was Glacial Erosion?

There are conflicting opinions on the amount of erosion accomplished by the continental ice sheets of North America. White (1972) believed that deep erosion in the central portions of the Canadian Shield created Hudson Bay and exposed the Precambrian rocks from beneath a younger Palaeozoic cover. He also proposed that a second major erosional zone runs along the Precambrian-Palaeozoic rock boundary, an area now occupied by the St Lawrence and upper Mackenzie Valleys, and by the Great Lakes, Winnipeg, Reindeer, Athabasca, Great Slave, and Great Bear lake basins. It has been suggested that the amount of sediment deposited on the continental margins of eastern North America is consistent with an average amount of erosion of at least 120 m in the Laurentide region (Bell and Laine 1985, White 1988).

The concept of deep glacial erosion has generally not been well received in the literature (Gravenor 1975, Sugden 1976, Andrews 1982). Many workers believe that only confined glacial flow, as in valleys and fiords, is able to accomplish large amounts of erosion, and that the unconfined flow of ice over much of the Shield modified the landscape to only a small degree. In support of this argument, critics have noted that most of the Precambrian Shield was probably exposed in the late Palaeozoic and early Mesozoic eras, long before the onset of Pleistocene glaciation. The glacial origin of Hudson Bay is disputed, and the tills deposited near the margins of the continental ice sheets do not contain large amounts of Shield material. The persistence of preglacial landforms on the Shield further belies the deep-erosion hypothesis, suggesting that glacial erosion did not substantially alter the preglacial bedrock morphology of the Shield. The glacial dispersion of a variety of rock types implies that resistant Shield rock was eroded by no more than 1 to 2 m in the last glacial stage. Although erosion was probably fairly substantial in the first couple of glacial stages, when the ice was able to remove deep, preglacially weathered material, most workers believe that the total amount of glacial erosion on the Shield during the Pleistocene was less than 20 m.

Landscape Models

Sugden (1978) calculated the morphology, dynamics, and basal temperature of the Laurentide ice sheet. Areas on the Shield where glacial erosion was dominant occur where his model suggests that the ice was warm-based. Erosion also dominated where the topography would have caused the ice flow to converge. More restricted or selective linear erosion occurred in the fiords on the eastern flanks of the ice sheet, where the model indicated that the ice was cold and protective on the plateaux and, because of ice convergence, warm and erosive in the pre-existing valleys. There was little or no glacial erosion on the High Arctic islands, where the model suggested that the ice was cold-based, and on uplands, where the flow was divergent (Fig. 6.12).

Sugden's model represents an interesting attempt to explain the erosional characteristics of the Laurentide ice sheet. It was based upon some questionable premises, however, including the assumption that the ice sheet had a single dome. The model stresses the importance of ice temperature, but other factors are also important. For example, it has been found that while systematic changes in the intensity of glacial erosion in the eastern Canadian Arctic can be related to basal ice temperatures, they are also partly the result of geological variations.

There have also been a few attempts to model glacial depositional zones

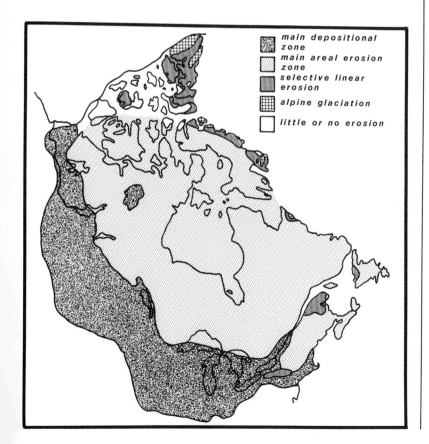

main depositional zone

main areal erosion zone

selective linear erosion

alpine glaciation

little or no erosion

Figure 6.12.
Generalized depositional and erosional zones in Canada (Sugden 1978).

at a variety of scales. Workers have identified three large landscape units created during Laurentide deglaciation (Flint 1971, Dyke and Prest 1987). Zone one, the most southerly unit beyond the margins of the Shield, has a thick and nearly continuous cover of glacial material, with large terminal and hummocky disintegration moraines. Long eskers and streamlined features occur in zone two, which includes most of the Shield and the northern Appalachians, but this area has far fewer moraines and generally much less glacial sediment than does zone one. The third zone in the final retreat areas in Labrador-New Québec and Keewatin has extensive Rogen moraine and a fairly thick and continuous cover of glacial material.

A simple model has been devised to account for the distribution of glacial depositional landforms (Fig. 6.13) (Sugden and John 1976). Erosion dominates in the central portions of an ice sheet, but deposition becomes progressively more important in the outer zone. Ice is still quite active in the inner part of this depositional zone. but not in the outer part, where it becomes sluggish or even stagnant. Basal debris in the active ice area is deposited as lodgement till and then streamlined. Deposition of ablation till and glaciofluvial sediments by melting ice dominates in the outer area, where debris becomes more common at all depths within the ice. A terminal moraine is produced at the ice margins, where the ice is moving slowly. Repeated glaciations produce wedges of sediment thickening towards the

Figure 6.13.
A possible sequence of glacial depositional landforms beneath part of the periphery of a mid-latitude ice sheet (Sugden and John 1976).

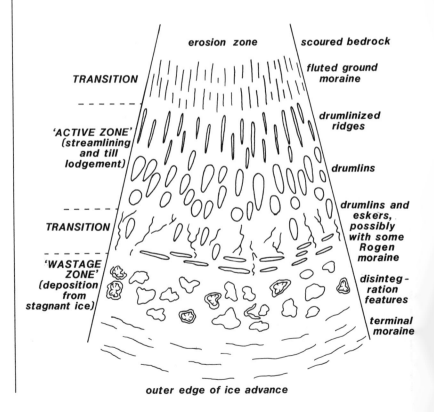

ice margins, superimposed on top of each other according to their relative age.

Moving from the central erosional zone where the bedrock is scoured, the model predicts the following sequence of landforms:

(1) fluted ground moraine in the erosional-depositional transition zone;

(2) drumlinized ridges and then drumlins in the active zone of streamlining and till lodgement;

(3) drumlins and eskers and possibly some Rogen moraine in the active depositional-stagnant depositional transition zone; and

(4) disintegration features, followed by terminal moraines in the stagnant ice zone.

The model emphasizes relationships between landforms in the glacial landscape, and provides a useful reference with which to compare field observations. Nevertheless, it remains to be adequately tested in the field, where numerous local factors influence the distribution of landforms. It has been reported, for instance, that there is some agreement between model predictions and the progression of landforms on either side of an ice divide in the Ungava Peninsula of northern Québec, although there are also substantial differences.

Glacial features are distributed in distinct patterns around the ice divides in Keewatin and Labrador-New Québec (Shilts et al. 1987, Aylsworth and Shilts 1989). Four roughly concentric zones have been distinguished around the Keewatin ice divide:

(1) The innermost zone, which extends for about 50 km on either side of the ice divide, consists of low till plains with almost no glaciofluvial or streamlined depositional features.

(2) The next zone is about 200 to 250 km wide, and is characterized by well-developed Rogen moraine and esker-outwash systems. The Rogen moraine is distributed in linear belts or trains radiating out from the dispersion centre, with featureless or drumlinized areas in between.

(3) The third zone is between 200 and 300 km in width. This zone has an intricate dendritic esker pattern, fairly continuous drift, and drumlin fields. The eskers are greater in number and have more tributaries than in zone 2, although they are generally smaller. Rogen moraine in this zone is restricted to small, widely scattered fields lying in depressions.

(4) The outermost zone consists of extensive outcrops of bedrock with little or no drift cover.

Less is known about the distribution of glacial features around the Labrador-New Québec ice divide. Nevertheless, recognizable patterns do exist, although they have been distorted in places by divergent ice flow around higher land. The innermost zone in Labrador-New Québec also lacks Rogen moraine and eskers, and the drift is generally quite thin. Streamlined features, including drumlins and crag and tails, are well developed in the second zone, along with Rogen moraine and long eskers. As in Keewatin, the drumlinized features occupy belts between fields of Rogen moraine.

The third zone contains complex esker systems and widely scattered fields of Rogen moraine, while the outermost zone has little drift.

The distribution of glacial landforms around the ice divides in Keewatin and Labrador-New Québec is therefore broadly similar. This suggests that changes in ice thickness, velocity, degree of stagnation, type of basal flow (compressive or extensive), and other aspects of ice dynamics are at least partly responsible for a regular arrangement of landforms around ice dispersal centres. The local topography and the geological character of the glacial bed, however, must also have played an important role. An increase in drift thickness and the presence of drumlins, eskers and Rogen moraine in the second and third zones in Labrador-New Québec, for example, partly reflect the occurrence of fairly easily eroded bedrock. Trains of Rogen moraine between the Keewatin ice divide and Hudson Bay can be traced back to outcrops of a distinctive late Archean granite. These outcrops probably provided a bouldery basal load for the ice, which in turn provided suitable conditions for the formation of Rogen moraine rather than the drumlins that are commonly found in the intervening boulder-free areas.

Periglaciation

The term 'periglacial' was originally introduced to refer to climatic and **geomorphic** conditions around the margins of Pleistocene ice sheets. Today, however, it is used for the processes and landforms of cold climates, regardless of any present or former proximity to glaciers. Permafrost, or perennially frozen ground, is characteristic of periglacial climates, but some periglacial activity is associated with frost action outside the permafrost domain. Changes in climate related to the advance and retreat of the ice sheets also brought into the periglacial domain huge areas that are now considered to lie outside it. Although few processes are restricted to cold environments, most operate quite differently in periglacial than in warmer regions, with marked differences in their frequency and intensity (Hamelin and Cook 1967, Brown 1970, French 1976, Lewkowicz 1989).

Periglacial Climates

There is no single periglacial climate; rather, there are several major kinds. The type of climate – including such factors as the relative importance of seasonal and **diurnal** temperature cycles, and the availability and distribution of moisture – is a vital factor in determining the development of periglacial landscapes (Figs. 1.12 and 1.13). Three main groups of periglacial environments are found in Canada: Arctic **tundra** or ice-free polar deserts; subarctic or northern forests; and alpine zones in the mid-latitudes.

The High Arctic has great seasonal differences in temperature, but very little daily variation. During the months of perpetual darkness, temperatures can be as low as −20 to −30°C. A thin surface layer, between about 0.3 and 1.5 m in depth, thaws for a few months in summer, when air temperatures are above zero. Precipitation is less than 100 mm in some areas, but evaporation is low and the frozen ground prevents the moisture from percolating deeply into the soil. The thin and discontinuous snowcover of the High Arctic allows deep penetration of the ground by frost.

Diurnal temperature variations are also weak in the subarctic and continental areas south of the treeline. The seasonal range of temperature is even greater than in the High Arctic, however, even though average annual temperatures are similar or only slightly higher. Summers are hotter and longer than in the High Arctic, with temperatures above zero for almost half the year; this allows the ground to thaw down to 2 or 3 m or more. Precipitation, which is also higher than in the High Arctic, falls mainly in summer. Fires occur quite often in the northern **boreal** forest, however, because the higher summer precipitation is offset by higher evaporation. Winter snowfall is also much greater than in the High Arctic, and plays an important role in determining the distribution of perennially frozen ground or permafrost. Snow accumulation in open areas prevents frost from deeply penetrating the ground. Although permafrost can form under trees where there is less snow and therefore deeper frost penetration, continuous and thick permafrost in the subarctic is probably a relict of more severe climatic conditions in the past.

Alpine periglacial climates exist above the treeline in the middle latitudes. The treeline is between 2,000 and 4,000 m in the Rockies, but it is close to sea level in northern Labrador-Ungava. There are seasonal and diurnal variations in precipitation and temperature in these environments, and temperatures frequently fluctuate around the freezing point. Nevertheless, permafrost is either absent or discontinuous in alpine climates, because snowfall is heavy and severe winters are rare.

Frost Action

Frost is the most widespread and important periglacial weathering mechanism. Three basic processes are involved:

(a) The contraction and cracking of ice-rich, frozen soil as it is cooled. These cracks tend to form a polygonal network that can be filled by ice, sand, or soil.

(b) The formation of **segregated** ice as a result of suction that draws water from unfrozen parts of the soil to a freezing surface. Segregated ice can form **lenses**, layers, or massive ground ice. Although the process is generally most effective in fine-grained materials, clays are less suitable than silts because of their low **permeability**. The formation of segregated ice in the seasonally thawed surface layer causes frost-heaving in the soil and the local doming or lifting of surface sediments; upward movement and tilting of stones, often until they stand on end; and the horizontal and vertical sorting of particles according to size.

(c) Pressures resulting from the approximately 9 per cent expansion of water upon freezing. Although it is unlikely that maximum pressures can be exerted under field conditions (Chapter 2), many types of rock deteriorate when they are exposed to temperature fluctuations around the freezing point. The evidence includes accumulations of shattered rock at the foot of cliffs and other very steep slopes (talus or scree), on mountain sides (block slopes), and on fairly level surfaces (block fields or **felsenmeere**). Felsenmeere are particularly well developed on the northeastern upland rim of the continent from Ellesmere Island to Labrador, but they are also found on many uplands and lowlands throughout Arctic Canada. Frost shattering (gelifraction) concentrated along linear fracture zones may also have excavated the small, vertically sided valleys or *ravins de gélivation* found throughout the Canadian Shield, although some workers believe that they were largely eroded by glacial meltwater. Bedrock has been heaved into domes, ridges, and conical mounds throughout the permafrost areas of northern Canada, as a result of the pressures generated by water freezing in **joints** and other cavities, and the attempted expulsion of water as freezing proceeds downwards from the surface. Because freeze-thaw cycles are not very frequent today in Arctic Canada (Fig. 2.5), explanations for the occurrence of extensive areas of frost-shattered **debris** have included arguments (a) that it has been produced very slowly over long periods of time; (b) that it was produced under more suitable conditions during the Pleistocene; and (c) that the importance of frost action is exaggerated by the insignificance of other weathering processes.

The freezing of water can also generate high cryostatic pressures. These pressures are produced when pockets of unfrozen soil are trapped between the downward freezing soil and the top of the perennially frozen ground. Unfrozen pockets are usually the result of differences in the rate of ground freezing according to variations in moisture content, grain size, or vegetation. Pressures generated by freezing in the seasonally frozen surface zone deform and displace sediments, producing aimlessly contorted structures or involutions in the soil.

Frost action was probably more intense in the mid-latitudes during Pleistocene glacial stages than it is in the high latitudes today. Seasonal differences in climate are greater in high latitudes, but the dominant diurnal cycle in the mid-latitudes could have produced between about 50 and 100 freeze-thaw cycles per year. Although these cycles would have been of short duration and their effect limited to shallow depths of soil, they would probably have generated much greater frost shattering, frost creep, and gelifluction (Chapter 3) than is experienced in the high latitudes today.

Permafrost

Permafrost – perennially frozen ground with temperatures persistently below 0°C – underlies about half of Canada. It can range from thin layers that have remained frozen from one winter to the next, to frozen ground hundreds of metres in thickness and thousands of years in age.

Two major zones are recognized in Canada (Fig. 7.1), both of which are basically controlled by climate. Permafrost underlies all the continuous

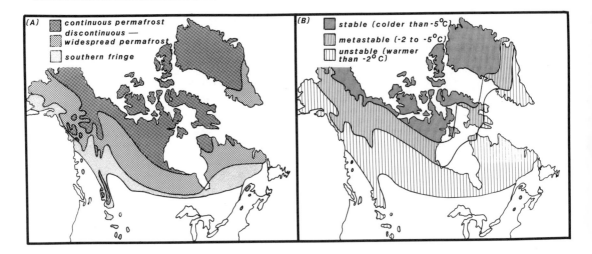

Figure 7.1
Distribution and stability of the permafrost zones in Canada (Harris 1986).

permafrost zone, with the exception of newly deposited sediments, as in the Mackenzie Delta, and large, deep bodies of water; it is quite extensive beneath the shallow waters off northwestern Canada and the Arctic islands. Frozen ground extends down to depths of 60-90 m at the southern boundary of the continuous zone, to more than 500 m in the Arctic Islands, over 700 m on Richards Island off the Mackenzie Delta, and possibly down to more than 1,000 m in the high areas of Baffin and Ellesmere Islands. The climate is generally less severe in the discontinuous zone to the south, where areas underlain by permafrost coexist with areas of unfrozen ground. The term 'sporadic permafrost' has been used to refer to the southern fringe of the discontinuous zone, where small, scattered islands of permafrost are largely restricted to peatlands (muskeg). The distribution of permafrost in the mountains of western Canada depends upon the occurrence of suitable combinations of altitude, aspect, snow cover, topography, vegetation, and wind, as well as latitudinal position.

The active layer above the permafrost is thawed in summer and only seasonally frozen. It extends down to depths ranging from about 10 cm on Ellesmere Island to 15 m in the higher areas of the Rockies. Taliks are unfrozen areas of variable shape within or below the permafrost, and also in some cases between the base of the active layer and the top of the permafrost (Fig. 7.2). In the discontinuous zone, unfrozen ground above and below the permafrost can be connected by chimney-like taliks, which perforate the frozen ground (Fig. 7.3). Taliks are open if they are in contact with the seasonally thawed active layer, and closed if they are completely surrounded by permafrost. Most open taliks are the result of heat from lakes and other bodies of standing water. Closed taliks develop for a number of reasons, including the draining of a lake, or ancient changes in climate.

Permafrost temperatures are usually lowest near the ground surface. Diurnal fluctuations occur down to perhaps 1 m, and seasonal fluctuations

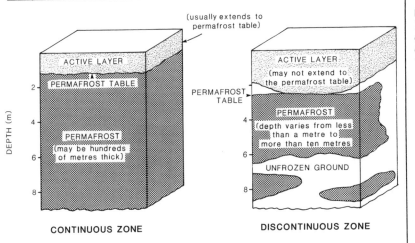

Figure 7.2.
Taliks and active layers in continuous and discontinuous permafrost (Brown 1970).

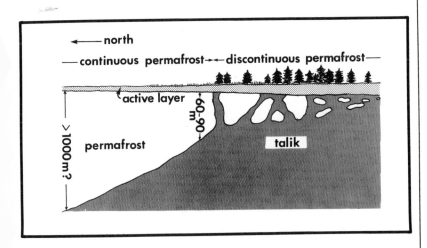

Figure 7.3.
Transect from north to south through the continuous and discontinuous permafrost zones of Canada (Brown 1970).

albeit with a considerable time lag, down to 15 m or more. Temperatures do not fluctuate at greater depths, but they gradually increase with depth, reaching the freezing point at the base of the permafrost.

Permafrost develops where the depth of freezing in winter exceeds the depth of thawing in summer. This creates a layer of frozen ground that persists through the summer. The frozen layer thickens with the addition of annual increments, until a rough balance or equilibrium is attained. Estimates can be made of the maximum depth of permafrost in an area based upon surface temperature and the amount of geothermal heat flowing from the interior of the Earth. Large errors may occur, however, because of the warming influence of large bodies of water, and the presence of deep, relict permafrost formed during the last glacial stage. It has been suggested that the depth of the permafrost at its southern limit is in general accordance with today's climate. There is evidence of patches of relict permafrost on the Labrador-Ungava Peninsula, however, and some permafrost in the western Canadian Arctic is at least early **Wisconsin** in age. Shoreline retreat and postglacial isostatic uplift have had an important effect on ground temperatures and permafrost depths in the High Arctic. On Cameron Island, for example, the permafrost is deepest where the land emerged earliest from the sea, and on Melville Island, the permafrost is twice as thick inland as in coastal areas. Numerous local factors also influence the occurrence and thickness of permafrost and the active layer in the discontinuous zone. They include the amount of snowcover, the type and density of the vegetation (FitzGibbon 1981), the orientation and gradient of the slopes, and the **albedo** and thermal conductivity of the soil and rock (Fig. 7.4).

Because of fairly short winters and the importance of diurnal temperature cycles, the climate of the mid-latitudes during Pleistocene glacial stages was probably less severe than the climate of polar and continental areas today. It is therefore unlikely that there would have been thick permafrost in the Pleistocene in the milder oceanic areas of the middle latitudes.

Figure 7.4.
The effect of vegetation and terrain on the distribution of permafrost in the peatland of the southern fringe of the discontinuous zone (Brown 1970).

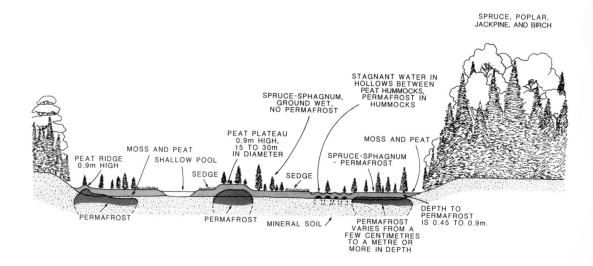

Ground Ice

The widespread occurrence of ice within frozen ground provides one of the greatest obstacles to northern development. It is found in many forms, including pore ice between the grains of soil, thin veins and films, large vertical wedges, horizontal sheets, and irregular masses tens of metres in thickness. It can also be formed in many ways, including the freezing of groundwater in a climatic cycle; **sublimation** and **condensation**; freezing of water percolating beneath peat and frozen ground; freezing of water in polygonal cracks; and burial of snowbanks, stagnant glacial ice, or drift ice on streams, lakes, or the sea. Some ice is diurnal, some is seasonal, and some is thousands of years in age.

There are thick exposures of ice in the western Arctic islands and in the coastal lowlands of the western Arctic mainland (Mackay 1972, Harry et al. 1988). Exploratory oil shot holes in the Mackenzie Delta region, for example, suggest that there is a 20 to 30 per cent chance of encountering icy sediments at any depth down to at least 40 m, and a 0 to 5 per cent chance of finding massive ice. On Richards Island in the Mackenzie Delta region, ice constitutes about 45 per cent of the volume of the upper few metres of permafrost. Similarly, areas underlain by shaley rock on eastern Melville Island contain between 50 and 70 per cent ice by volume at depths of between 1.5 and 2 m, and between 30 and 40 per cent down to depths of about 10 m (French et al. 1986).

Ice Wedges

Ice wedges are typically V-shaped bodies of ground ice extending downwards 3 to 4 m into the permafrost, but in parts of the western Arctic they can be more than twice this size (Mackay 1989). While active or growing ice wedges are most abundant in areas of continuous permafrost, there are inactive forms in the discontinuous zone. Ice wedges are thought to develop in cracks that form when the ice in frozen ground contracts, at temperatures below − 15 to − 20°C (Fig. 7.5). Moisture flows into the cracks in the spring and freezes, preventing the crack from closing. The growing ice

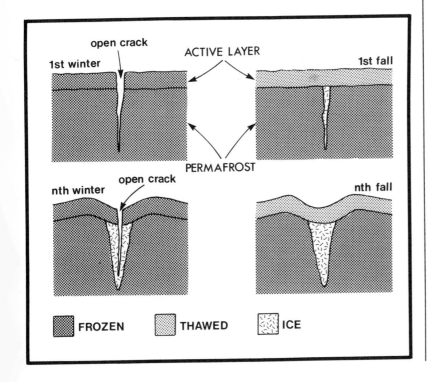

Figure 7.5.
The origin of ice wedges (Lachenbruch 1962).

wedge provides a plane of weakness that tends to re-open under stress in winter, allowing new ice layers to be deposited. In the arid polar desert, where the moisture to form ice is lacking, 'sand' wedges form as the cracks are filled with windblown sand or other material.

Large numbers of wedges often form a polygonal pattern on the ground surface. Unlike most 'periglacial' features, ice and sand wedges and polygons develop only in perennially frozen ground. There are a few other features that occur only in permafrost regions, including thermokarst and pingos, but they are difficult to identify. Polygons may therefore provide the only reliable proof of the former existence of permafrost conditions in an area (Fig. 7.6).

After the ice melts, the form of a wedge may be preserved by debris that falls or is washed into the hole. Ice-wedge casts and fossil polygonal patterns have been identified in southern Ontario, the St Lawrence Lowlands, and the Maritime provinces, and ancient sand wedges have been recognized in the drier parts of the western Interior Plains. The distribution of ice- and sand-wedge casts suggests that during the Pleistocene, the permafrost zone in North America was much narrower than in Europe. This was probably because of the greater compression of the climatic zones in North America, owing to the more southerly extent of the ice sheet. The presence of large glacial lakes around much of the periphery of the retreating Wisconsin ice sheet (Fig. 5.8) also limited the area that was exposed to periglacial conditions, and the time available for the development of complete or mature polygonal networks.

Ice-cored Mounds

Many types of mound in the Canadian north have an ice core. They are members of a diverse family of features that differ according to origin, nature of the contained ice, size, number, and longevity.

Pingos

Pingos are ice-cored hills ranging from a few metres to more than 60 m in height (Mackay 1966). Some are symmetrically conical, while others, particularly those growing in abandoned stream channels, are quite elongated. Examination of the interiors of pingos suggests that their cores consist of pore ice and varying amounts of **intrusive** ice, wedge ice, dilation-crack ice, and lenses of segregated ice in silts and sands. This icy core may be buried beneath 1 to 10 m of overburden. As the ice core melts, the decay of a pingo is marked first by the formation of a summit crater and possibly a small lake, and then by the collapse of the structure to form a shallow, rimmed depression. More than 1,440 pingos have been identified on the coastal plains of the western Arctic, but they also occur in the Yukon, to the north and northeast of Great Bear Lake, on several Arctic Islands, and on the northern Ungava Peninsula of Québec. Some pingo-like features have also been identified on the floor of the Beaufort Sea.

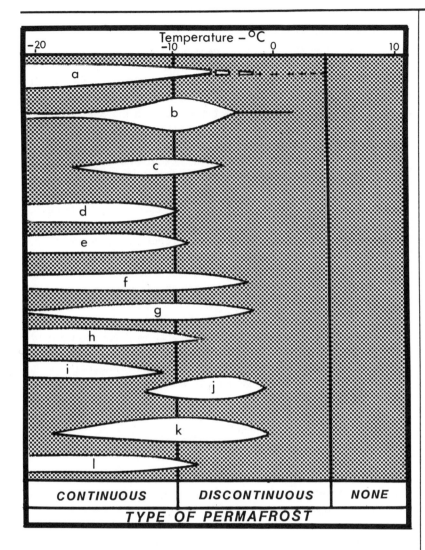

Figure 7.6.
*The distribution of perigla-
cial landforms (Harris
1981). The thickness of the
shapes represents the frequency of occurrence.*

a Permafrost

b Thermokarst

c Ice-wedge polygons - peat

d Ice-wedge polygons - mineral

e Non-sorted polygons

f Sorted polygons

g Open system pingos

h Closed system pingos

i Holocene blockfields

j Peat plateaux

k Palsas

l Earth hummocks

The distinction is often made between open- (hydraulic) and closed-
(hydrostatic) system pingos, depending on the source of the water that
forms the ice core. Open-system pingos are particularly numerous in the

Figure 7.7.
The origin of open and closed pingos (French 1976, based on the work of F. Muller and J. R. Mackay).

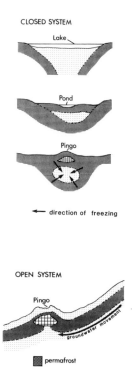

CLOSED SYSTEM

Lake

Pond

Pingo

← direction of freezing

OPEN SYSTEM

Pingo

groundwater movement

■ permafrost

□ unfrozen ground

▦ ice

central Yukon and the adjacent parts of Alaska. They are found in areas of thin or discontinuous permafrost, where water can drain into the ground and circulate within taliks or beneath the permafrost. Pingos are formed where this water rises to the surface and freezes, forcing up the overlying sediments (Fig. 7.7). Open-system pingos tend to develop as isolated features or in small groups in valley bottoms and on the central or lower portions of slopes, where enough pressure can be generated to force the water upwards. For reasons that remain unknown, virtually all the open-system pingos in northwestern North America are in unglaciated regions, on the south- or, especially, southeastern-facing slopes of valleys. A possible exception has recently been discovered on the Ungava Peninsula in Québec, however, where two pingos have developed in very fractured rock near the bottom of a glaciated valley (Seppala 1988).

Closed-system pingos are found in areas of continuous permafrost, often where there is little relief. There are more than one thousand on the Pleistocene coastal plain of the Mackenzie Delta, and others occur on the modern delta, on the coastal plain of the Yukon, and on western Victoria and southern Banks Islands. A few have also been identified in Keewatin, on Baffin Island, and possibly on the floor of the Beaufort Sea. Single pingos on the Mackenzie Delta usually develop in shallow ponds formed by the rapid drainage of a large lake, but several pingos in valleys on Banks Island, and one on an alluvial fan on Baffin Island, may be the result of the closure of taliks under shifting stream systems. In the Mackenzie area, it is thought that as the lakes shrink in size, closed systems of unfrozen material are trapped between the freezing surface layers and the advancing permafrost. Pore water expelled before the advancing freezing front then leads to the growth of ice, the updoming of the surface, and the formation of pingos (Fig. 7.7).

Some pingos seem to be of neither the open nor the closed variety. There are, for example, more than 100 small, circular pingos aligned in two parallel rows along the summit of Prince Patrick Island, far from any lakes, and in an area of thick, continuous permafrost. They are of unknown origin, but may be related to some deep geological structure. Another group of pingos includes some elongated esker-like forms.

Palsas

Palsas are low hills of peat, 1 to 10 m in height, containing a permafrost core with numerous thin ice lenses and bands. Most numerous near the southern fringes of the discontinuous permafrost zone, particularly in the Hudson Bay Lowlands of Manitoba and Ontario, they have also been identified in the Albertan Foothills and in the Yukon and Northwest Territories. Palsas and palsa-like forms are not restricted to the discontinuous permafrost zone, however, and they have been reported on the mainland and islands of Arctic Canada. Peat plateaux, which are larger features formed by the coalescence of palsas, are common in northern Alberta and northeastern British Columbia, and on the Shield towards the northern limit of the discontinuous zone.

Palsas are frost-heaved peat accumulations resulting from the formation of segregated or injection ice in the underlying soil, often where snow cover is thin or absent. Their association with bogs in the muskeg zone of discontinuous permafrost reflects the insulating properties of peaty organic material, and the protection it affords to the underlying permafrost. The eventual decay and collapse of palsas produce a thermokarst landscape consisting of pits, hollows, and small lakes.

Frost Blisters

Several other small mounds contain ice cores or ice lenses. Seasonal frost blisters, for example, are widespread in Arctic and subarctic regions. These ice-cored mounds may reach a few metres in height, and range in diameter from a few metres up to 70 m. They differ from palsas in their rate of development and mode of formation; they grow to their maximum size in a single winter, and decay and disappear within a few months or years. They are the result of springwater that, under high pressure, freezes and uplifts soil and organic sediments during the period of winter freezeback.

Thermokarst

Thermokarst is hummocky, irregular relief produced by the melting of ice-rich permafrost and subsidence of the ground. The term reflects the similarity in form to karst terrain in limestone regions (Chapter 10). Thermokarst can be the result of an increase in the depth of the active layer because of climatic warming, destruction of insulating vegetation by fire, natural changes in vegetation, loss of water, shifting stream channels, erosion of the active layer, or the constructional and transportational activities of man. Recent climatic warming has probably been an insignificant factor in the western Canadian Arctic, although increasing concentrations of greenhouse gases in the atmosphere could result in extensive thermokarst development over the next 50 to 100 years.

There has been a considerable increase in human activity in the north in the last few decades, largely associated with exploration for oil and gas. This has involved excavation, drilling, and transportation over land, and the construction of production and processing facilities, settlements, roads, airfields, and power and water supplies. Controlling or preventing the local disturbance of permafrost by human activity is one of the major challenges for northern development.

Thermokarst features include collapsed pingos, mounds, thaw lakes, sinkholes, beaded drainage (where blocks of ground ice melt and locally enlarge a stream channel), and polygonal and linear troughs produced by the melting of ice-wedge polygons. Well-developed thermokarst terrain is not widespread in Canada, however, although there are thermokarst depressions and related features on Ellef Ringnes, Banks, and southern Victoria Islands, in the northern Yukon, and in the Mackenzie Delta and Tuktoyaktuk coastal areas. Thermokarst is probably of less significance in mountainous regions, in the eastern Arctic, and on the Shield, where much of the surface consists of resistant rock.

The most common thermokarst feature in Canada is probably the shallow, rounded thaw lake or depression, which is widespread on the Arctic mainland and on the southern Arctic islands. They may form initially through the random melting of massive ground or wedge ice, followed by subsidence of the ground and accumulation of water. Lakes grow as the surrounding permafrost melts and adjacent lakes coalesce. The depressions may be eventually filled in with silt and organic matter, and by gelifluction and other forms of mass movement from the sides, but they can also drain suddenly if an adjacent lake depression expands into them. The cycle, involving the growth and decay of a lake, can be completed within a few thousand years.

Oriented Lakes

Many thaw lakes tend to be elliptical in plan shape, with their long axes oriented in a common direction, at right angles to the prevailing winds. This orientation appears to reflect the occurrence of zones of maximum current velocity, littoral drift, and erosion. Nevertheless, a full explanation of the relationship between wind direction and lake orientation awaits further study of such factors as wave-current systems, lake ice, and thermal erosion, and their effect on thaw rates and the transportation and **deposition** of sediment.

Oriented lakes have been reported on the coastal plain of the Yukon and Mackenzie District, in the interior of the Yukon Territory, on the western coast of Baffin Island, on the southwestern coast of Banks Island, and elsewhere in Arctic Canada. Permafrost may not be essential for the formation of oriented lakes, however, as they are found in a variety of environments around the world.

Patterned Ground

The ground surface in periglacial regions is often characterized by a variety of cells, mounds, and ridges distributed in fairly regular geometric patterns (Zoltai and Tarnocai 1981). Patterned ground is not restricted to cold climatic regions, but the patterns are usually more conspicuous in periglacial areas than in other environments. Periglacial patterns normally develop in polar and subpolar regions, where there is a large annual range in temperature. Smaller forms are produced by diurnal frost cycles in alpine areas, however, and there are even several active types on the Avalon Peninsula, the southern coast of Newfoundland, and the southeastern coast of Cape Breton Island, where there is abundant moisture and freeze-thaw activity.

The main geometric shapes, including circles, polygons, nets, steps, and stripes, can be produced in a sorted or nonsorted form. Sorted patterns involve the separation of the larger material from the smaller, whereas nonsorted patterns are generally defined by changes in elevation or vegetation. There is usually a transition from polygons, circles, and nets, on essentially flat surfaces, into steps and then into stripes, as slopes steepen and mass movement becomes more significant. Slopes that are greater than about 30° are usually too steep for patterned ground to develop.

Sorted circles usually consist of fine material in the centre, with larger stones in the outer rim. The debris island is a type of sorted circle composed of fine material surrounded by blocks and boulders on steep, debris-covered slopes. The stony border is absent on nonsorted circles, which are typically slightly dome-shaped and bordered by vegetation. The borders of large, nonsorted polygons in permafrost regions are often the site of ice wedges. Two types of ice-wedge polygon can be distinguished. Polygons with low centres have marginal ridges, and may hold standing water in summer. Polygons with high centres can develop from low-centred polygons, although they also develop as a primary form. Nonsorted polygons range from forms that are less than 1 m in diameter up to large ice-wedge or tundra polygons that are often between 5 and 40 m across. Sorted polygons have a border of coarser material and are usually only found on flat surfaces, whereas nonsorted polygons can develop on quite steep slopes. Nets are patterns that are neither circular nor polygonal, although they too can be sorted or nonsorted. Earth hummocks, or thufur, are a hummocky, nonsorted type of net consisting of small, vegetated domes about 0.5 m in height. Terrace-like features or steps develop on slopes from circles, polygons, and nets. The step riser is lined with stones in the sorted form, but only by vegetation in the generally much smaller nonsorted form. Steps that have become lobed or drawn out downslope are known as garlands. Nonsorted stripes consist of lines of vegetation with intervening areas of bare soil running downslope, whereas sorted stripes are alternations of coarse and fine material.

The origin of patterned ground remains unclear, although many theories have been proposed to account for the numerous types which have been identified. It is thought that similar kinds of patterned ground can be created by different processes, while the same processes can produce a variety of forms. For example, round to elongate soil patches or mud boils, which are a common type of nonsorted circle on the Shield and in the western Arctic, are formed when semi-liquid mud erupts into the surface zone. Several workers have suggested that this is the result of cryostatic pressures (Fig. 7.8), although in some areas other processes appear to be responsible for their development. Hummocky structures can be produced in several ways (Mackay 1980). Some workers believe that at least some of the widespread earth hummocks in the Yukon, the Mackenzie Valley, the

Figure 7.8.
The possible origin of mud boils. Cryostatic pressure exerted on the water-saturated fine material, which is the last to freeze, is relieved by its injection into the surface layers (French 1976).

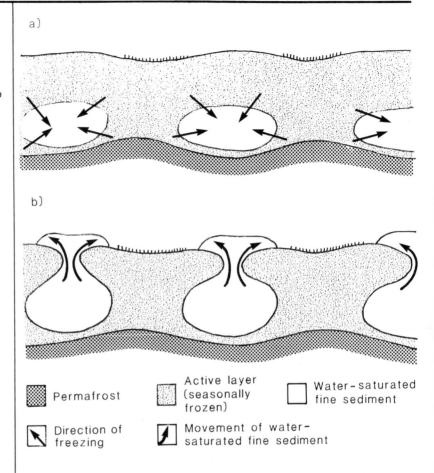

western and central Arctic, and the Arctic islands are the result of the displacement of wet, unfrozen pockets of ground by cryostatic pressures. It has also been suggested, however, that they are produced by freeze-thaw associated with the growth and decay of ice lenses at the top and bottom of the active layer. Sorted circles may be formed in two stages. Upfreezing first produces a surface layer of dense, coarse material with finer material below. The weight of the coarse material then deforms the underlying sediment, causing plugs of fine material to rise to the surface at certain points; these plugs provide the centres for sorted circles. Cracking of the ground appears to be important for the development of polygonal shapes, but not circular or step forms. Needle ice may be responsible for the development of stripes in the Rocky Mountains.

Nivation

Nivation involves the erosion of a hillside by frost, gelifluction, and slopewash beneath and, particularly, at the fluctuating edge of lingering banks of snow. Its effectiveness depends upon the thickness of the snowbank and whether there is permafrost beneath it. Shallow nivation hollows develop where there is abundant meltwater for intense frost action, and the debris can be removed by gelifluction and slopewash.

Protalus ramparts are ridges of generally coarse debris that form at the lower edge of permanent or semi-permanent snowbanks. One type is built by frost-rived rock falling from cliffs and then sliding down snow surfaces, while a similar ridge of fine material is produced by gelifluction beneath snowbanks.

Slopes

Slopes in periglacial regions are generally similar to those in other climatic environments. There are some significant differences, however, which are attributable to such factors as the importance of frost action, the lack of vegetation, and the presence of frozen ground. (Periglacial mass movements on slopes are discussed in Chapter 3.)

Slope Deposits and Landforms

Frost-creep and gelifluction (Chapter 3; Fig. 3.7) deposits can be in the form of sheets, lobes, terraces or benches, and streams. Gelifluction sheets produce smooth terrain with gradients of only 1 to 3°. They are characteristic of the High Arctic, where the absence of vegetation allows slow mass movement to operate fairly uniformly. Further south in the tundra and forest tundra areas, the presence of vegetation favours movement in the form of tongue-like lobes or terraces. Lobes and terraces can be turf- or stone-banked (stone garlands), depending on the presence and nature of the material concentrated at their downstream ends. They occur in most periglacial regions, but are probably most common in mountainous areas and in other rugged terrain. Gelifluction lobes are active today in the southern Rocky Mountains, with average surface movements of about 0.61 cm yr^{-1} (Smith 1987b). Gelifluction can also have a pronounced linear form. Active block streams, for example, are slow-moving, elongated bodies of loose rock found in valleys, or forming narrow, linear deposits down steep slopes (Fig. 3.2 j).

Some slope deposits (*grèzes litées*) are rhythmically bedded, with alternating layers of fine and coarse material ranging from silts up to rock particles 10 cm or more in length. The deposits are thought to be the result of seasonal or possibly diurnal freeze-thaw cycles, with slopewash producing bedding and crude sorting of the sediments. Although bedded slope deposits have been recognized on Banks Island, they appear to be quite rare in high latitudes, which do not experience the frequent freeze-thaw cycles of the mid-latitudes.

Figure 7.9.
Typical slope profiles of periglacial regions (French 1976).

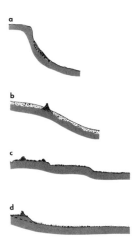

Rock Glaciers

Rock glaciers are lobes or tongues of angular rock and fine material in the highlands of polar, subpolar, and mid-latitude regions (Luckman and Crockett 1978, Johnson 1978, 1984b). They range from several hundred metres to more than a kilometre in length, and up to 50 m in thickness. Rock glaciers are particularly numerous in the valleys of the St Elias Mountains in the southwestern Yukon, and in the Rockies. In Jasper National Park, for example, one hundred and nineteen rock glaciers have been identified within an area of 4,632 km².

The occurrence of furrows, crevasses, ridges, and other surface features on rock glaciers testifies to their present or former movement. The mechanisms are poorly understood, but movement may be related to the presence of a glacial ice core, **interstitial** ice, or some form of creep. Rock glaciers can be formed in several ways. Tongue-shaped forms develop from debris-covered glaciers, and they usually occur in, or issue from, cirques. There is, however, some dispute over the origin, and indeed the existence, of some nonglacial or periglacial lobate forms fed by debris from talus or avalanche slopes.

Slope Profiles

There are all kinds of slope profile in periglacial climates, and no particular type of slope can be considered to be typical of these areas. Nevertheless, for convenience one can recognize four common types of slope profile in periglacial regions (Fig. 7.9) (French 1976):

(a) Some profiles consist of a very steep cliff or free face above a concave debris slope. Frost action and rockfalls cause cliff retreat, and debris accumulates as talus or scree, usually with a gradient of between 30 and 40°. Talus deposits, however, usually mantle rock surfaces and are rarely as thick as they seem. Gelifluction and **slopewash** operate on the gentler slopes below the talus.

(b) Slopes mantled in frost-shattered and gelifluction debris have fairly smooth profiles with gradients usually between 10 and 30°. Residual hillside **tors** may project above the debris on the upper portion of the valley sides.

(c) Nivation and frost action cut gently sloping **cryo**planation (altiplanation) terraces in the middle and upper portions of some slopes. Although these terraces are in rock, they are generally covered by a thin veneer of gelifluction and **sheetwash** debris. Cryoplanation terraces have been identified in the mountains of the Yukon (Lauriol and Godbout 1988), in the Arctic Archipelago, in Keewatin, and in central Labrador, but they do not appear to be very common in the eastern Arctic.

(d) Cryopediments are very gently concave erosional surfaces, which are usually cut into the base of valley-side slopes. Frost action, mass movement, and **rillwash** operate on the valley sides, and gelifluction and sheetwash transport the debris across the cryopediment. There are often residual hilltop or summit tors surrounded by gentle slopes on the **inter-fluves** and other high areas of profile types (c) and (d).

Many authorities believe that slopes in periglacial regions gradually become smoother and flatter as erosion is concentrated on the higher sections with deposition on the lower. Relief could be reduced either through parallel retreat of frost-rived rock scarps and the extension of the gentle slopes below, or, as is happening in sedimentary rocks in much of northern Canada, by the gradual inundation of steep cliffs by talus or debris slopes.

Wind Action

Direct wind action is thought to be of fairly minor importance in periglacial regions today, although it affects soil temperatures and snow cover, and therefore influences the work of other processes and mechanisms, including gelifluction, nivation, and **fluvial** action. Winds were very strong near the ice margins during the Pleistocene, however, when ice advance caused the latitudinal compression of the climatic zones, and the development of high-pressure anticyclonic conditions. Effective wind action was also helped at that time by the presence of bare or sparsely vegetated fine-grained sediments, which were produced by frost action or deposited at the margins of the ice sheets.

Vegetation inhibits the winnowing out and transportation of fine material by the wind. Wind **deflation** is therefore probably most important in the polar deserts, although it can function in tundra regions on outwash trains (Chapter 6) and recent deltas, wherever the vegetational cover is sparse (Good and Bryant 1985). Deflation of fine material can leave a surface lag deposit or desert pavement of coarse material, and it can also produce shallow depressions, or blowouts.

Loess

Much of the world's most productive farmland is on loess: a buff-coloured quartz-silt that is usually derived from the grinding down of rock by glaciers, or by frost in cold, high mountains. Wind deflation of **till** plains or outwash plains over long periods of time resulted in the deposition of thick accumulations of loess over neighbouring regions. The greatest deposits in North America are in the north-central United States, where they are arranged in a broad east-west belt south of the glacial limit. Small pockets of loess have been mapped in Canada in the western Interior Plains, the Peace River area (Fig. 7.10), and southern Ontario (Geological Society of America 1952), and there are discontinuous patches, less than 0.5 m in thickness, throughout the Banff-Jasper area of the Rockies. There are also modern loess deposits in the Yukon, and it is being deposited around the margins of the polar desert today. As the study of loess in Canada progresses, it is probable that other areas will be identified where loess lies in small, shallow deposits, or is an important constituent of the soil (Smalley 1984).

Figure 7.10.
Distribution of loess in western Canada (Geological Society of America 1952).

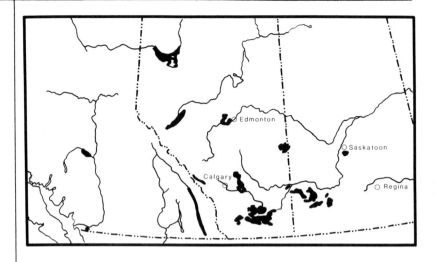

Sand Dunes and Wind Abrasion

There are isolated, and in some cases extensive, areas of wind-blown sand in most parts of Canada, although they have generally been stabilized by vegetation today (Fig. 7.11) (David 1977, 1981). These areas are usually found on, or in the vicinity of, deltas, outwash, or other sandy **glaciolacustrine**, **glaciomarine**, or **glaciofluvial** deposits. Sandy hills or **dunes** are especially common in Alberta and Saskatchewan, particularly around Lake Athabasca, where they are still active or moving in some areas.

It is thought that there are two main kinds of dune in Canada. Parabolic dunes, which have a crescentic plan shape, with the horns facing towards the wind, formed where the sand was moist, whereas transverse dune ridges, at right angles to the wind, developed in dry sand. Subsequent transformation and elongation of parabolic dunes has produced long, narrow ridges of sand in several areas.

Sand dunes record palaeo-wind directions and other aspects of climatic change since the retreat of the Laurentide ice sheet. The presence and form of parabolic sand dunes in Alberta, for example, have been used to determine former wind conditions in areas that were marginal to the ice sheet (Odynsky 1958). The widespread occurrence of parabolic forms across Canada suggests that, contrary to popular belief, most dunes did not develop in dry, desert environments, but rather in areas with sufficient moisture to encourage the growth of vegetation.

Another indication of the intensity of both modern and former wind action is the widespread occurrence of striations, or scratch marks, on rock outcrops, and ventifacts, or pebbles worn down or grooved on their exposed sides by wind-borne abrasives. These abrasives may include blowing snow, which attains a hardness similar to that of sand grains at a temperature of $-45°C$.

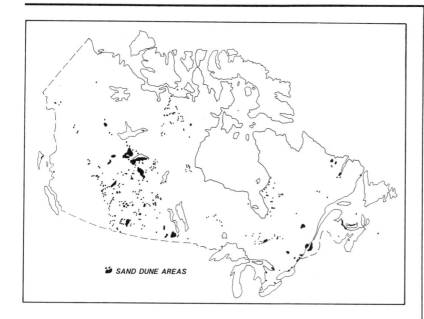

SAND DUNE AREAS

Figure 7.11.
Sand dunes in Canada (David 1977).

Streams and Valleys

It is becoming increasingly evident that running water plays an important role in shaping the landscapes of cold regions. Small streams in the Arctic can flow only for a few months during the summer, when temperatures are above freezing. Nevertheless, they accomplish a great deal of erosional and transportational work in that short time, relative to the work of other geomorphic agents. Even on semi-arid Axel Heiberg Island in the High Arctic, running water from occasionally heavy summer rain is thought to be the most important geomorphic agent today. Valley deepening and widening are continuing in the High Arctic, and rillwash and sheetwash are important on a smaller scale.

Fluvial Regimes

The flow in all streams in northern Canada is dominated by the rapid melting of snow and ice in spring or summer. Between 25 and 75 per cent of the total annual runoff is concentrated within a few days, and the streams therefore tend to flood. Most of the sediment load is also carried during this short breakup period.

The melt period in the Canadian Arctic occurs in late June or early July. Runoff then gradually decreases through the rest of the summer, although there may be sudden increases for short periods in response to summer storms. Nival flow regimes of this type are termed Arctic if there is no winter flow, and subarctic if some flow is maintained during the winter. A proglacial regime is one in which snow and icefields provide meltwater throughout the warmer periods of summer. Maximum runoff is usually in late July or early August, and is less pronounced than in nival regimes. Flood flows are reduced or damped in a wetland or muskeg regime because of the water-holding capacity of grassy tundra and its high resistance to runoff.

Fluvial Processes

Fluvial processes in cold regions are strongly influenced by snow and ice, and the presence of permafrost at shallow depths (Woo and Sauriol 1980). Most of the stream flow in the High Arctic takes place when the valleys are choked with snow. Stream flow cuts channels or tunnels in the snow during the melt period. These unstable channels shift vertically and laterally in the snow and ice cover, exposing a broad zone to fluvial activity. Meltwater that is impounded in valleys behind large snow drifts can generate high discharges when the latter are eventually breached. Ice jams also develop in the narrower sections of some large streams, and **ice floes** can groove and scour their beds and banks. Ice frozen to the bottom of streams, however, provides some protection during the initial breakup period. Lateral migration of streams in cold climates is assisted by **thermo-erosion** of ice wedges and ice-rich material in the banks (Chapter 3), although in some cases the presence of permafrost can increase the resistance of stream banks to erosion.

Asymmetrical Valleys

Some valleys in periglacial regions are asymmetrical in cross-section, particularly where they run east-west. North-facing slopes are steeper in the central Canadian Arctic, but the steeper slopes are oriented in other directions on Banks Island and in the extreme High Arctic.

The most plausible explanation for steeper north-facing slopes involves gelifluction and lateral stream erosion. Gelifluction is greater on south and southwest-facing slopes, which receive more solar radiation and therefore have deeper active layers than slopes facing in other directions. When valleys run approximately east-west, the debris delivered by gelifluction from the south-facing slopes forces the streams to the opposite banks, where they undercut and steepen the north-facing slopes. A rather different situation occurs at higher latitudes. On Banks Island, the steeper slopes of asymmetrical valleys face towards the west and southwest. Possibly because of westerly winds and small daily variations in the inclination of the sun at high latitudes, these slopes are drier and cooler than others, and have shallower active layers. Gelifluction and nivation are therefore most active on the east-facing slopes, and the debris they produce forces the streams to undercut and steepen the west-facing slopes. In the Caribou Hills in the Northwest Territories, conditions change from Arctic to subarctic from the top of the plateau down to the base of the hills. Orientation of the steepest valley side slopes in this area depends upon their altitude. Where the climate is more severe at higher elevations, the steepest slopes face to the north, but where the climate is milder in the lower valley zone, the steepest slopes face to the south. This reversal emphasizes the relationship between slope orientation, microclimate, and basal stream activity in determining the occurrence and nature of asymmetric valleys in periglacial regions (Kennedy and Melton 1972).

String Bogs

String bogs (patterned fens) in muskeg areas consist of alternations of thin, string-like strips or ridges of peat and vegetation with shallow, linear depressions and ponds. They are found in subarctic Canada from the lower Mackenzie to central Labrador-Ungava, and are particularly prominent in the Hudson Bay area. The ridges are up to about 1.5 m in height, 1 to 3 m in width, and tens of metres in length. The linear patterns are normally oriented transversely to the gentle regional slope. The peat ridges contain ice lenses for at least part of the year, and in some areas they may include true palsas. Explanations for the origin of string bogs have included the effects of gelifluction, frost thrusting of ridges from adjacent ponds, differential frost heaving, growth of ice lenses, differential thawing of permafrost, and hydrological and botanical factors; however, none of these theories has yet proved satisfactory.

8 | Rivers

Streams are important components of the hydrological cycle, by which water is continuously circulated, along a variety of paths, between the land, the oceans, and the atmosphere (Fig. 8.1). When water evaporates from the sea, or from any other surface, some of the water vapour **condenses** and precipitation falls onto the land. Some of this water is stored in pools and as other forms of surface moisture, and is eventually evaporated back into the atmosphere. Water also infiltrates into the ground, adding to the soil moisture, and **percolates** downwards to the underlying **groundwater**. This water, which can be stored for long periods, may eventually be drawn back to the surface, where evaporation from the ground and **transpiration** from vegetation return it to the atmosphere. Groundwater also seeps and flows into surface streams, adding to the water delivered by surface runoff.

The presence of streams reflects the tendency of water running off the land to concentrate in channels. Different parts of a stream tend to function as primarily **erosional**, transportational, or **depositional** zones. Hillslopes and numerous small channels feed water and **sediment** into the drainage system in the upper zone. This zone passes downstream into a sediment transport zone containing one or a small number of channels, where the sediment entering at the upstream ends is approximately equal to the amount leaving at the downstream ends. The lowest zone is a sediment sink, where sediment from a single channel is deposited in fans, deltas, or **alluvial** plains, or in deeper water. Although this simple model is a convenient generalization of reality, it should be noted that sediment erosion, transportation, and deposition occur to some extent in all parts of a stream.

The Effect of Glaciation

Most Canadian drainage systems (Fig. 8.2) have been **deranged** by glaciation. The rivers are therefore geologically very young, and the country

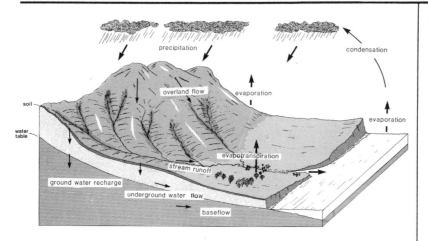

Figure 8.1.
The hydrological cycle.

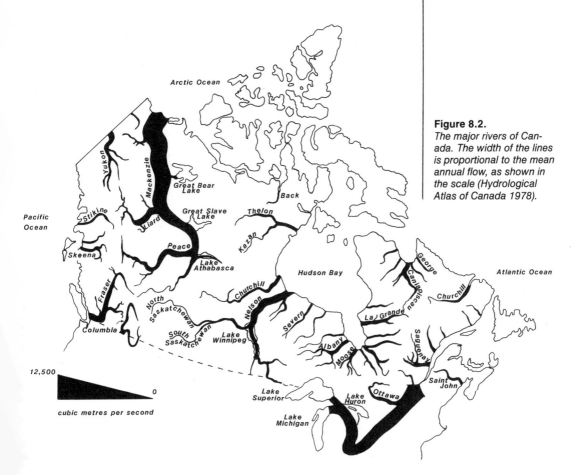

Figure 8.2.
The major rivers of Canada. The width of the lines is proportional to the mean annual flow, as shown in the scale (Hydrological Atlas of Canada 1978).

has no integrated drainage network comparable to the almost complete **dendritic** pattern that presumably existed before the onset of glaciation. The gradients of streams were altered by glaciation, and their sediments continue to be captured in numerous glacially scoured or **debris**-dammed lake basins, especially on the Shield and around its perimeter. The St Lawrence, for example, drains a very large area, but because of the Great Lakes it carries little sediment. Other large rivers, including the Athabasca, Peace, and Slave, also lose much of their sediment in lake basins along their courses. Furthermore, while the sediment in most streams in unglaciated regions becomes progressively finer downstream, the situation is usually more complex in Canada because of the influx of glacial material along stream courses (Church and Slaymaker 1989). This material often contains rocks and minerals that were carried by the ice from other areas and from other drainage basins.

Many preglacial valleys were partially or entirely filled with glacial deposits. Some of these valleys were obliterated, while others were re-excavated by postglacial streams. Most of the Assiniboine and Qu'Appelle Valleys in Manitoba and Saskatchewan (Klassen 1972), and parts of the modern drainage system in Alberta, were inherited from preglacial channels, but some sections were buried in glacial sediment and abandoned (Fig. 8.3). Borings and seismic methods have also revealed the existence of a fossilized valley network blocked by glacial deposits around Sherbrooke in the Eastern Townships of Québec (Clément and Poulin 1975).

Figure 8.3.
Bedrock channels in Alberta and Saskatchewan (from Farvolden 1963 and Christiansen 1967).

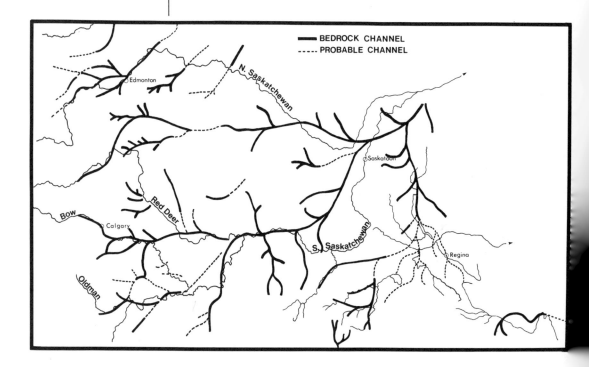

Meltwater flowing from ice sheets and draining from glacial lakes also cut new channels into glacial deposits, and in places into bedrock, and some still carry permanent streams. In the mid-continental regions, valleys that originated as spillways draining glacial lakes tend to be deep, narrow trenches, whereas the former meltwater channels are shallow and wide. Spillways in the central Interior Plains were rapidly cut by outburst floods from glacial lakes. The influx of large amounts of water from a lake into one at lower elevation then caused that lake to overflow and deeply incise its outlet. In this way, the sudden drainage of a lake triggered other interconnected lakes to drain and cut deep spillways (Kehew and Lord 1989). The Ouimet Canyon, between Thunder Bay and Nipigon in Ontario, is a spectacular example of a spillway that now contains no stream. The narrow canyon, at least 100 m deep, was probably formed by **proglacial** or subglacial drainage along a north-south trending fault (Kor and Teller 1986).

The Drainage Basin

Precipitation is absorbed into the ground at a rate that is determined by such factors as the vegetational cover and the texture and structure of the soil. Water runs down the surface of a slope when the rainfall intensity is greater than the capacity of the surface to absorb it. The runoff increases downslope as more and more water is added to it, forming stream channels where the flow becomes sufficient to overcome the resistance of the surface to erosion. Erosion may initially form a series of parallel rills running downslope, but as deeper, wider rills capture smaller adjacent ones, stream channels form. These gullies and channels then lengthen as they grow headwards, undercutting the slopes at their rear. The headward growth of the channels may be assisted by the incorporation of underground pipes or tunnels formed by groundwater flowing downslope through the upper levels of nearly saturated soil; pipe collapse was responsible for the formation of most of the major gulley systems in the Milk River Canyon, in the semi-arid **badland** areas of southeastern Alberta (Barendregt and Ongley 1977).

Headward erosion and extension of stream channels can result in the capture of the headwaters of one stream by another. The headward or easterly growth of the Annapolis River in Nova Scotia captured the flow of several northerly flowing streams, making them tributary to it. The former valleys of these streams, north of the Annapolis River, are now marked by a series of dry valleys or gaps, including the Digby Gut (Fig. 8.4). In the Tertiary, many streams flowing from the Rockies across the Interior Plains and the Shield were probably tributaries of a major river that occupied the basin of Hudson Bay, on the floor of which the former river pattern can still be detected. The northeasterly flowing rivers are represented today by the Nelson and Saskatchewan Rivers, but the others were captured by the headward extension of the Mackenzie and Mississippi-Missouri Rivers. The Liard, Hay, Peace, and Athabasca Rivers, for example, are now tributaries of the Mackenzie (Fig. 8.2). It is believed that

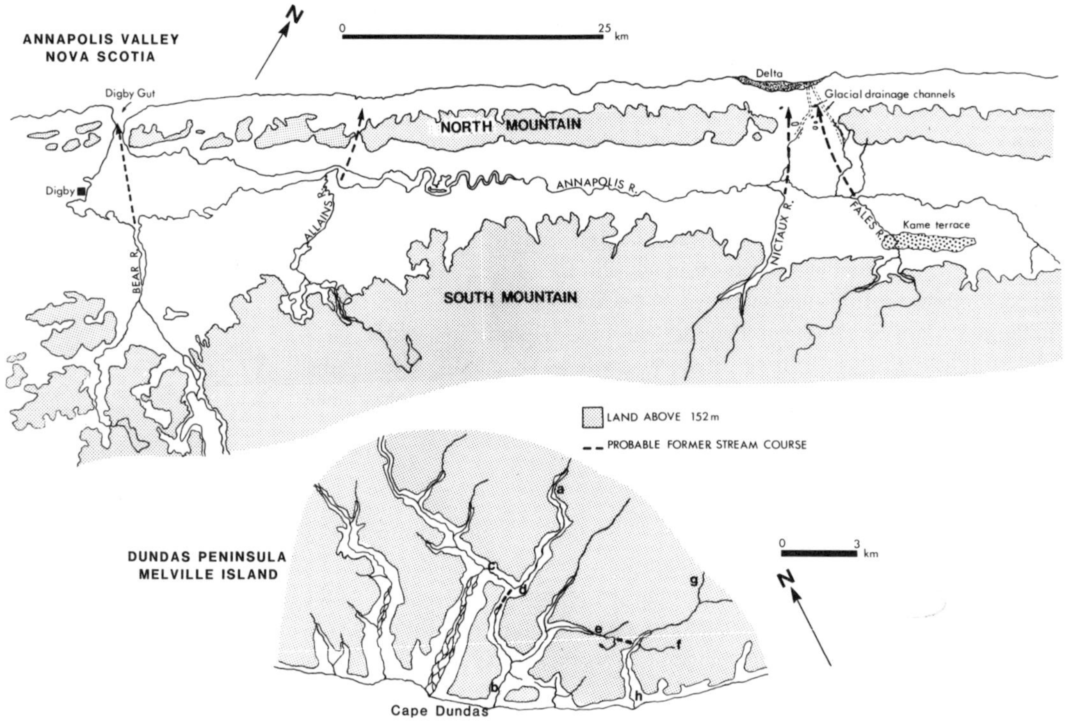

Figure 8.4.
Stream capture in the Annapolis Valley of Nova Scotia, and on an unnamed stream on Melville Island in the Arctic. The first example shows the capture of the Bear, Allains, Nictaux, and Fales Rivers owing to the eastward extension of the Annapolis River. These streams formerly flowed through North Mountain into the Bay of Fundy. The stream on the Dundas Peninsula, Melville Island, used to flow along ab, but it has been captured by the extension of an adjacent stream along cd. Another branch of the stream (ef) may have been captured by stream gh.

stream capture (piracy) during the Tertiary also played an important role in the formation of the Arctic Archipelago. Streams flowing northwards into the Arctic Ocean when sea level was lower may have been captured by other streams growing along the crustal weakness that is now occupied by the Parry Channel, and the Archipelago may have been created when these valleys were drowned by a rise in sea level. South- and southeasterly-flowing rivers in southeastern Canada may also have been captured by the headward extension of the predecessors of the St Lawrence (Fig. 8.5).

A drainage basin is the area drained by a stream and its tributaries. The drainage pattern that develops within the basin is strongly influenced by the nature of the underlying material, including the grain size of the sediments and the structure and lithology of the rock (Fig. 8.6). It has been suggested, for example, that drainage networks in the western Interior Plains and in southern Ontario are controlled by **tectonic** factors, as reflected in the orientation of **joints** and other structural elements. On the other hand, the orientation of some dry valleys in southern Alberta closely corresponds to the direction of the strong west-southwest Chinook winds. These ephemeral stream valleys (coulees) are narrow, fairly steep, and essentially parallel to each other. They are usually found on the windward slopes of the valleys and are tributary to the perennial streams. They are thought to have originated as surface furrows formed by wind-driven snow

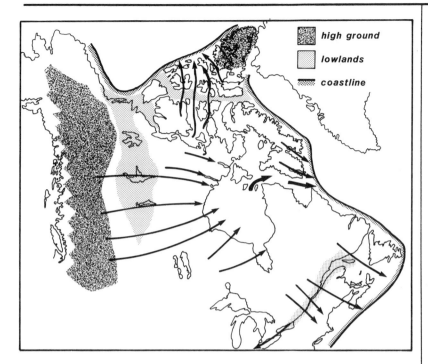

high ground

lowlands

coastline

Figure 8.5.
Possible Tertiary drainage in Canada (Bird 1972).

Figure 8.6.
Some types of drainage pattern (Ritter 1978).

DENDRITIC
Horizontally bedded sedimentary rocks or resistant crystalline rocks, with a gentle regional slope.

ANNULAR
Structural domes and basins.

RADIAL
Domes, including volcanoes.

TRELLIS
Dipping or folded rocks and areas with parallel fractures.

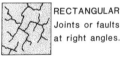
RECTANGULAR
Joints or faults at right angles.

PARALLEL
Moderate to steep slopes or parallel, elongated landforms.

CONTORTED
Contorted metamorphic rocks.

MULTIBASINAL
Karstic regions, permafrost, and hummocky deposits.

and rain, which were enlarged by mass movement and runoff following heavy rainfall (Beaty 1975). In the Ogilvie Mountains in the Yukon, stream channels are most numerous on the shaded north- and northeasterly-facing slopes, where prolonged snow cover provides a long period of meltwater flow (Liebling and Scherp 1983).

Following the original work of Horton, several schemes have been devised to describe the organization of a stream network within a drainage basin. Workers have defined a stream hierarchy, in which the most minor tributaries have the lowest order, and the main or trunk river the highest. The number, length, and gradient of the streams and the area and relief of their drainage basins have been found to be related to stream order, according to the 'laws' of drainage composition (Fig. 8.7). Most, but not all, drainage basins obey these morphometric laws, although it is not clear what they actually represent. Workers disagree over whether they are the result of an orderly or predestined development of drainage networks, or statistical relationships generated by random processes of development.

Stream Flow

Streams can be classified according to how frequently they flow. Ephemeral streams are usually dry, but they carry water during and immediately after rainfall. Intermittent streams are dry for part of the year and wet at other times, and are fed by groundwater when the **water table** is high enough.

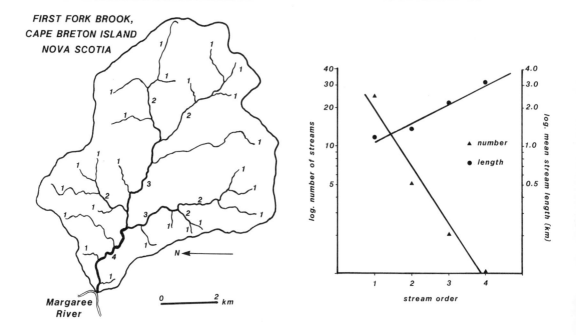

Figure 8.7.
The 'laws' of drainage composition using Strahler's system for stream ordering. Streams of a given order only have tributaries of lower order. First-order streams therefore have no tributaries, whereas second-order streams (or stream segments) have only first-order tributaries. A second-order stream begins where two first-orders meet, and ends where it meets another second-order stream, so starting a third-order stream.

Perennial streams flow all year round, and receive a fairly steady supply of groundwater.

The discharge of a stream at a point along its course is the amount of water passing through the channel cross-section at that point during a specified interval of time. It is given by:

$$Q = wdv$$

where Q is the discharge (m³ sec⁻¹); w is the width and d the depth of the water (m); and v is the velocity of flow (m sec⁻¹). Canada's largest rivers, the St Lawrence and the Mackenzie, have mean annual discharges at their mouths of 10,100 and 9,910 cubic metres per second, respectively (Fig. 8.2).

Once repeated measurements have been made at a site to establish the relationship between the height or stage of the stream surface (gauge height) and the amount of water flowing in the channel, future estimates of discharge at that site can be based simply on the water level. This information is used to determine the magnitude and frequency of floods, by drawing hydrographs showing changes in discharge through time, and flow-duration graphs showing how often a given discharge is equalled or exceeded.

Nearly all Canadian streams have extremely variable levels of flow, although natural storage of water in muskeg or lakes, or the building of dams by man, tends to smooth out seasonal variations. High discharges on most Canadian streams are generated by snow melt; in the spring in eastern

Canada and, at least for glacially fed streams and small streams in high mountains, in summer in the west. Discharge then gradually declines in the fall and winter, reaching a minimum in early spring. Exceptions to this pattern occur in the high Arctic, where the major rivers are completely frozen in winter, and in coastal British Columbia, where discharges are high in winter and low in summer.

Water moves in either laminar or turbulent flow. Laminar flow takes place along parallel individual paths, in layers that seem to slide over each other with no significant mixing. Groundwater moves in laminar flow, but occurs only in natural streams near the bed and banks. Stream flow is largely turbulent. Turbulent flow involves the chaotic movement of water, considerable mixing, irregular paths of fluid flow, and eddies superimposed on the main forward movement.

The dimensionless Reynolds Number (Re) is usually used to define the transition from laminar to turbulent flow:

$$Re = p\frac{vR}{u}$$

where p is the density and u is the viscosity of the water. R, the hydraulic radius of the channel, is given by

$$R = \frac{A}{P}$$

where A is the cross-sectional area of the channel (w × d); and P is the wetted perimeter, the edge of the cross section where the water is in contact with the channel (w + 2d).

The Reynolds Number is high for turbulent flow (greater than 2,000) and low for laminar (less than 500). The zone of maximum turbulence is near the bottom of a stream, but velocity is usually highest just below the water surface in the centre of the stream. There are two types of turbulent flow. Streaming flow is most common, but shooting flow, which is much more erosive, replaces it at high velocity. The Froude Number (Fr) defines the turbulent nature of a stream:

$$Fr = \frac{v}{\sqrt{gd}}$$

where g is gravitational acceleration. Streaming or tranquil flow occurs when the Froude number is less than one, whereas shooting flow occurs at values greater than one.

The Movement of Sediment

Most of the energy in a stream is converted into heat by turbulent mixing and, to a much lesser extent, by friction at the channel sides and base. Only a very small proportion is therefore available to pick up and move sediment. Nevertheless, streams do transport very large amounts of sediment as solutional, suspended, and bed loads. Each year, Canadian rivers carry approximately 300 million tons of sediment into the oceans and into the United States. The amount of sediment carried in suspension each year to the mouth of the Fraser River in British Columbia, for instance, would form a conical pile about 300 m high (*Hydrological Atlas of Canada* 1978). Because of such factors as long winters and the absence of agriculture over large areas, however, Canadian streams generally carry less sediment than those in the United States, or in most other parts of the world.

The movement of sediment varies in streams from place to place and from time to time. Weak weathered rock and glacial and other **unconsolidated** deposits can provide large amounts of sediment from fairly small areas. For example, badlands constitute only about 1.8 per cent of the Red Deer drainage basin in southern Alberta, yet they are the source of almost all the river's suspended sediment.

The proportion of dissolved matter to solid particles is usually high where groundwater makes the major contribution to stream flow, and in streams charged with organic acids supplied by decaying vegetation in bogs, marshes, and swamps. The solutional load also tends to be high in streams that flow over limestones, dolomites, and evaporites (Chapter 10), or over **tills** with a high carbonate content, and low in areas of Shield bedrock (Fig. 8.8). The Athabasca and North Saskatchewan Rivers drain large areas of the southern Rockies. Much more of their drainage areas consists of carbonate than sulphate-bearing rocks, but because the latter are eroded much more rapidly, sulphates contribute between about 30 and 55 per cent of the dissolved load in both rivers, according to their discharge (Drake and Ford 1974). The amount of dissolved carbonate and sulphate in the Athabasca and North Saskatchewan Rivers shows a close correlation with the amount of runoff, all three variables attaining their maxima in June and July. Nevertheless, large amounts of dissolved sediment are also transported during periods of lower discharge.

The suspended load consists of solid particles of clay, silt, and **colloidal material** carried in suspension in streams. While their concentration is greatest near the bottom, they can be supported by turbulence in the water above the layer of laminar flow. The suspended load is measured at stations throughout Canada, although the network is still rather sparse and the records are short in most regions. The amount of solid sediment moved by a stream depends upon such factors as stream discharge, gradient, velocity, the roughness of the bed, channel morphology, and the nature of the grains. Suspended sediment concentrations are high in the western Interior Plains and in other places with similar alluvial formations; in fast-flowing streams in mountainous regions; in southern Canada where the runoff period is longer than in the north; and in areas of agriculture and forestry, where natural vegetation has been disturbed or removed by man (Fig. 8.9) (Stichling 1974). The amount of suspended sediment is generally low in winter,

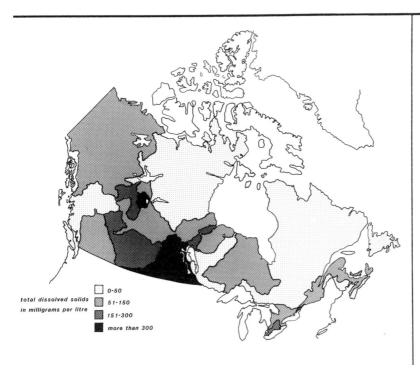

Figure 8.8
*Total dissolved solids in Canadian rivers (*Hydrological Atlas of Canada *1978).*

total dissolved solids
in milligrams per litre

☐ 0-50
▨ 51-150
▧ 151-300
■ more than 300

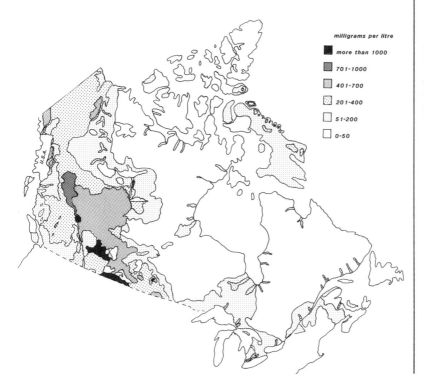

milligrams per litre

■ more than 1000
▨ 701-1000
▧ 401-700
▨ 201-400
☐ 51-200
☐ 0-50

Figure 8.9.
*Suspended sediment concentrations in Canadian rivers (*Hydrological Atlas of Canada *1978). About 60 per cent of the country has concentrations of less than 50 mg l⁻¹.*

Figure 8.10
Mean discharge (line graph) and total suspended sediment load (shaded bar graph) per month for four streams in 1983.

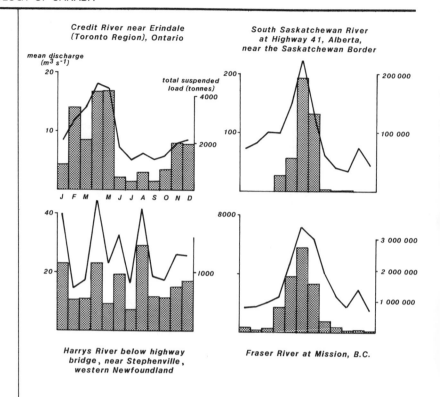

when the drainage basins are frozen or snow-covered, and high during the floods of spring or summer (Fig. 8.10). In southern Ontario, for instance, half the annual suspended load is carried in March and April, when 60 per cent of the annual extreme floods occur (Dickinson et al. 1975).

Bed load consists of sand and coarser particles sliding, rolling (traction load), or jumping (saltation load) along the floor of a stream in the lower layers of laminar flow. Lower velocities are required to move loose sand on a stream bed than to move either coarser gravel or finer silt (Fig. 8.11). This is because of, on the one hand, the size and weight of the coarse material, and on the other, the cohesiveness and smoothness of stream bottoms composed of fine particles. The velocity required to move particles at the bottom of a stream is also affected by such factors as grain density, bottom slope, and the homogeneity of the bottom sediments. The bed load in a stream is much more difficult to determine than the suspended load, and it is measured at only a few stations in Canada. It has been estimated, however, that it constitutes between about 5 and 20 per cent of the total sediment carried by streams.

'Stream capacity' refers to the maximum amount of sediment of a particular size that a stream can carry as bed load, whereas its 'competency' is the maximum size of the largest grain that it can carry as bed load. The

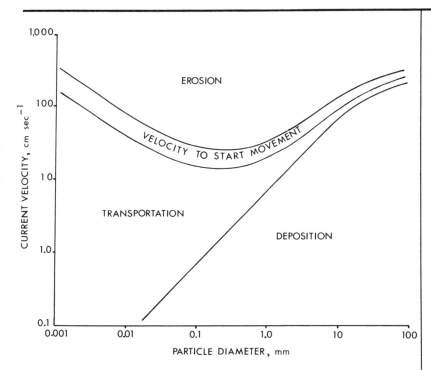

Figure 8.11.
*The relationship between sediment grain diameter and **entrainment** velocity in water (Hjulstrom, in Morisawa 1968).*

competency of a stream changes according to changes in discharge along its course, and at any one station through time. A large proportion of the sediment transported in gravel streams is moved during fairly infrequent periods of high discharge; on the Elbow River in southwestern Alberta, gravel on the stream bed is in transport for only about 30 days per year.

There appears to be a rough relationship between current velocity, bed-forms, and the movement of sediment in streams. Ripples develop on a stream bed when the Froude Number is much lower than one, corresponding to low stream velocity and a placid water surface. The small amount of sediment that is moved under these conditions is restricted to single grains on the stream bed. **Dunes** are larger features, which develop at higher Froude Numbers than do ripples. Although the Froude Number is still less than one, the water surface is less placid, and eddies form on the lee sides of the dunes. Grains now move over the dunes in groups, and the dunes advance downstream. At higher stream velocities, the stream bed becomes planar and there is considerable movement of sediment. As the Froude Number becomes greater than one, at even higher velocities, standing waves and then **antidunes** develop, with a great increase in the amount of sediment moving along the bed (Fig. 8.12).

The relationship between bed forms and flow conditions in the sand and gravel streams of Alberta does not fit the simple ripple-dune classification. Dunes are common on the sandy bed of the lower Red Deer River and ripples are formed at low water stages. Froude Numbers at all flow stages, however, are well below the critical level necessary for antidunes (Neill

Figure 8.12.
The relationship between stream bedforms, stream velocity, and sediment movement (Ritter 1978).

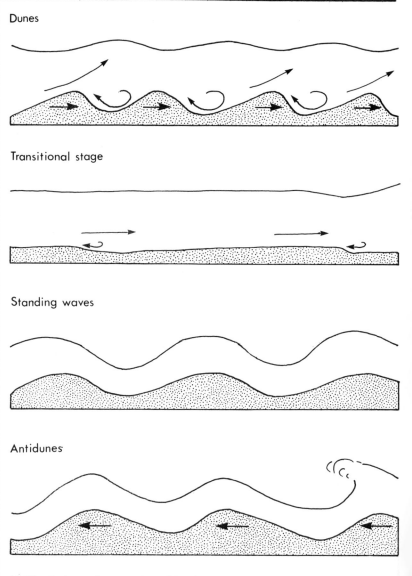

Dunes

Transitional stage

Standing waves

Antidunes

1969). At least three scales of bedforms can be identified on the Beaver River, ranging in size from small ripples with wavelengths of about 15 cm, through dunes with wavelengths of about 3 m, up to wave-like forms with wavelengths of a hundred metres or more.

Stream Erosion and Deposition

A stream cuts downwards when it is able to carry more sediment than is being supplied to it, and **aggrades** when more sediment is being supplied to it than it can carry. In the first case, the stream bed will be lowered; in the second, it will tend to rise.

Stream erosion is accomplished by **corrosion**, corrasion, and cavitation. Corrosion is the result of a reaction between water and the rocks with which it is in contact, whereas corrasion is the mechanical wearing away of the rock. The most important corrasional process is usually abrasion, which is caused by the grinding or impact of material carried by the stream against its bed and banks, although the impact of the water itself may remove some loose particles. Corrasion along a stream bed often polishes rock surfaces or excavates potholes in them. Cavitation involves the generation of shock pressures by the collapse of air bubbles. It is most effective at waterfalls, in rapids, and in other places where water velocity is very high.

Particles that have been picked up and carried by a stream are deposited when the current is unable to transport them any further. A decrease in the capacity and/or competency of a stream may therefore be the result of such factors as a decline in discharge, velocity, or channel gradient, or an increase in the size or amount of the sediment that is being supplied. Streams with sandy beds respond quickly to changes in discharge. Changes in the elevation of the stream bed at two bridge constrictions on the Beaver River in east-central Alberta, for example, mirrored changes in the water level during the passage of a flood (Neill 1965). The bed was lowered by several metres during the peak of the flood and then refilled, with a lag of a few days, as the water level fell (Fig. 8.13).

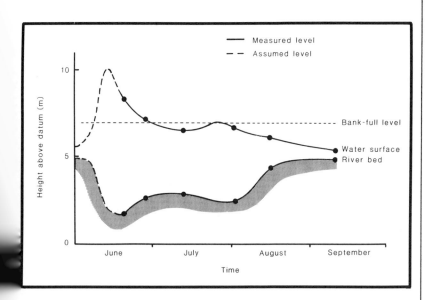

Figure 8.13.
Changes in the depth of the Beaver River, Alberta, with changes in discharge. Note erosion of the stream bed during the June flood (Neill 1965).

Channel Morphology

The distinction must be made between stream channels cut in bedrock and streams flowing in **fluvial** sediment. Only in the latter case are channels able to adjust their size, shape, pattern, and slope comparatively rapidly in response to changes in discharge and sediment. At any point along the course of a stream flowing in sediment, the width and depth of the channel and the velocity of flow change in response to changes in discharge (Q) through time. These relationships can be approximated by the simple power functions

$$w = a Q^b$$
$$d = c Q^f$$
$$v = k Q^m$$

where the constants $b + f + m = 1$; and $a \times c \times k = 1$.

There is a very strong relationship between channel width, cross-sectional area, and two-year flood discharge in Albertan streams with gravel beds, but the relationship between depth and discharge is a little weaker, and that between velocity and discharge is much weaker. On the other hand, the channels of the Beatton River (in northeastern British Columbia), which are in silty sand with some gravel, and Nigel Creek (in a small basin of the Rocky Mountains of west-central Alberta) accommodate increases in discharge with approximately equal increases in width, depth, and flow velocity (Fig. 8.14) (Hickin and Nanson 1975, Knighton 1976). **Cohesive** sediments often make it difficult for lowland streams to increase in width as their discharge increases, but rapid increases are possible in non-cohesive materials.

The width, depth, and velocity of streams also change along their courses in response to changes in discharge. Although these **spatial** changes can also be expressed by simple power functions, there are some important differences with the **temporal** changes that occur at a single site. Channel width increases more rapidly downstream with increasing discharge than at any one site on a stream, but changes in depth with discharge are similar. Flow velocity, however, changes more rapidly with discharge at a site, because the downstream increase in discharge is countered by a decrease in slope. At-site changes in hydraulic geometry with discharge on two tributaries of the Lillooet River in British Columbia are similar to those reported from elsewhere, but downstream changes are significantly different. It may be that in these and similar streams in Canada, there has not been enough time since the end of glaciation for an integrated equilibrium system to be established throughout the whole drainage system. Different sections of these streams have therefore not yet adjusted to each other.

The type and amount of sediment carried in a stream channel also have an important effect on its morphology. Streams carrying predominantly coarse material, such as those on the North Slope of the Yukon, tend to be wide and shallow, while streams with fine sediment loads, such as the lower Red Deer in southern Alberta, have narrow and deep channels. The relative amounts of sediment carried as bed load and suspended load also influence the cross-sectional shape of stream channels.

The effect of vegetational type and density on the morphology of stream channels remains to be determined. Nevertheless, it seems logical to assume that vegetation affects such factors as the channel's resistance to stream flow, the strength of stream banks, the formation of log jams, and the deposition of sediment on bars (Hickin 1984). The channel of a small stream in Québec, for instance, is deep and narrow where it runs through pasture, and shallower and wider where alder is growing on at least one of its banks (Bergeron and Roy 1985). It should also be noted that in Canada, unlike the warmer parts of the United States, channel width equilibria are upset by the effects of ice jams and ice scour.

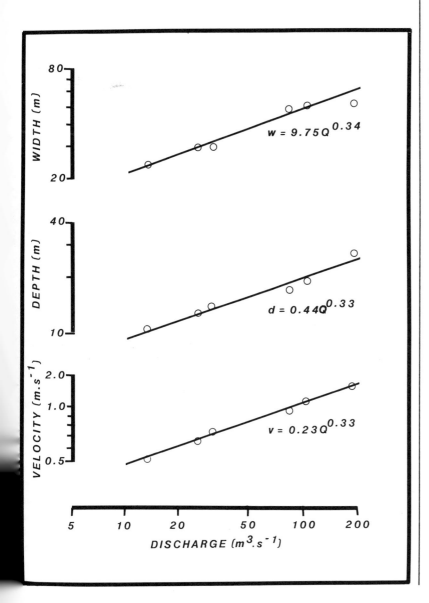

Figure 8.14.
The hydraulic geometry of the Beatton River, north-eastern British Columbia (Hickin and Nanson 1975).

Channel Pattern

'Channel pattern' refers to the overall appearance of a stream, particularly whether it is straight, meandering, or braided. These categories are not exclusive, as meandering streams can contain braided reaches, for example, and braided streams can contain meandering channels. Irregular patterns are those that cannot be included in the other classes.

Straight

Several rivers in the western Interior Plains are essentially straight for most of their length, and some of the tributaries of the St Lawrence are straight for considerable distances. Nevertheless, few streams are perfectly straight for any great distance, and even those that are have many of the same channel features as meandering streams. These include bars that alternate downstream, extending out from one side of the channel then the other, so that the thalweg, or the line joining the deepest parts of the channel, meanders between the straight banks of the stream. The stream floor undulates between areas of deeper water or pools opposite the bars, and shallower riffles, or shoals, about midway between successive pools (Fig. 8.15). Riffles, both in straight and in meandering streams, are separated by a distance equal to about five to seven times the channel width.

Figure 8.15.
Bars, riffles, and pools on a straight stream.

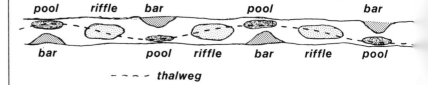

Meandering

Meandering is an inherent property of running water, but although a number of explanations have been advanced to account for it, none is entirely satisfactory. When water flows in a winding channel, centrifugal force directs the current to the outside of the bends, raising the water level. The **hydraulic head** associated with the higher water level intensifies helical flow, a spiral or corkscrew-like secondary flow that is superimposed on the general downstream movement of the water. This produces a strong downward flow of water and erosion at the outer bank. Water moves along the bottom of the stream to the inside of the curve, promoting deposition and the formation of a point bar. The occurrence of an eddy and more slowly moving water on the inside of the curve further assists in the formation of the bar (Fig. 8.16). The zone of maximum erosion or undercutting on the outside of a meander, and the corresponding point of maximum deposition on the inside, are offset a little downstream of the meander axis. Meanders therefore tend to slowly migrate laterally (outwards) and downstream, although movement is primarily lateral or downvalley on some streams.

Channel migration can produce meander scrolls, an alternating series of sandy ridges and silt-clay troughs (swales, sloughs) on the inner or convex banks of meanders (Fig. 8.16). Meander scrolls provide a historical record of the direction and growth of meanders on the extremely sinuous Beatton River in northeastern British Columbia. The scroll pattern suggests that there are marked changes in the growth of meanders when the ratio of the radius of channel curvature (Fig. 8.16) to channel width falls to about 2.11 (Hickin 1974). When this critical curvature ratio is attained, the zone of maximum lateral erosion may be shifted from the normal position, which is opposite but slightly downstream of the point bar, further downstream or even onto the upstream limb of the meander. On the very tightly curved bends of the Squamish River in British Columbia, for example, deposition rather than erosion is occurring on the outer or concave bank, and channel migration has stopped or has even been reversed in direction. Postglacial isostatic tilting has also influenced meanders on the Beatton River, superimposing an easterly bias on their general downstream migration.

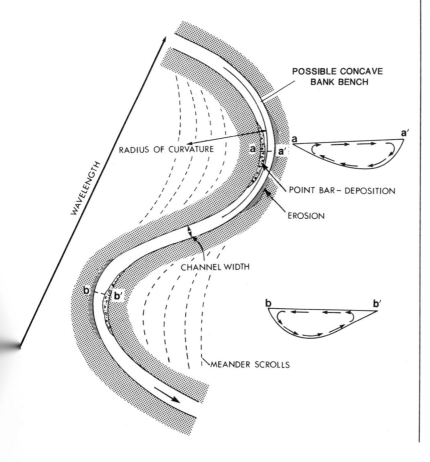

Figure 8.16.
Some characteristics of stream meanders. The position of a concave bank bench is shown as well, although such formations tend to develop on stream bends that are much tighter than those depicted here.

Migration rates of 10-15 m yr^{-1} or more have been recorded on the banks of some meanders on large streams, while other bends on the same streams are stable or moving only very slowly. The North Saskatchewan River laterally migrated by more than 100 m in places during only a few weeks of high floods in 1965, but the average rates of meander migration on eighteen streams in Alberta and British Columbia are between about 0.57 and 7.26 m yr^{-1}. The volume of sediment that is eroded by meander migration increases with the size of the stream and decreases with the grain size of the material at the base of the outer bank. Migration is a markedly discontinuous process, and the actual rate for any particular bend depends upon its channel curvature ratio. The rate of meander migration on streams in western Canada is at a maximum when the curvature ratio is between 2.0 and 3.0 (Hickin and Nanson 1975, Nanson and Hickin 1986).

Some meandering streams sweep across wide floodplains, rarely coming into contact with the valley sides. The meanders on many streams, however, are confined, impinging against the sides of valleys, or against bluffs of valley infill. Confined meanders are commonly incised or deeply **entrenched** in alluvium or bedrock, which may exert a strong influence on their form and rate of development. There are numerous Canadian examples, including all the rivers crossing the Coastal Plain of the Yukon, and the South Saskatchewan, Athabasca and other rivers crossing the western Interior Plains.

Meandering is an attempt to minimize the work done by a stream at some point along its course by lengthening the channel and decreasing its gradient. Although extremely sinuous meanders can develop, the path of a stream may be shortened in a number of ways. A meander loop can be gradually abandoned by chute cutoff, when a stream re-occupies the trough of an old meander scroll, diverting more and more of its flow from the main channel. Abandonment is much more abrupt, however, when two meanders intercept, or new channels are cut during periods of overbank flooding. The channel of the Muskwa River in northeastern British Columbia was shortened by about 4.1 km during a 36-hour storm in 1977. More than 2 million tons of sediment derived from the excavation of the new channel were deposited at the river mouth in the Fort Nelson River (Sherstone 1983). Abandoned channel loops become oxbow lakes, which are eventually filled in with **backswamp** clay (Fig. 8.17). Artificial cutoffs and stream straightening are frequently used in Canada to facilitate bridge-building, flood control, and the protection of agricultural land, but they have not always produced the desired or expected results.

Braided

Braided streams consist of numerous dividing and reuniting channels, with intervening sand or gravel bars or islands (D.G. Smith 1973, N.D. Smith 1974). One or more channels are usually dominant in large braided streams (Fig. 8.18). Changes in the position of islands and channels are often rapid,

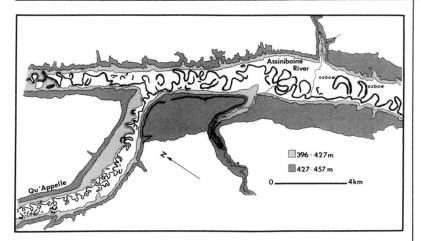

Figure 8.17.
Meanders and oxbow lakes on underfit streams in western Manitoba. The Assiniboine and Qu'Appelle Valleys formed as spillways and meltwater channels within partly excavated ancestral systems. Elevations given in metres above sea level.

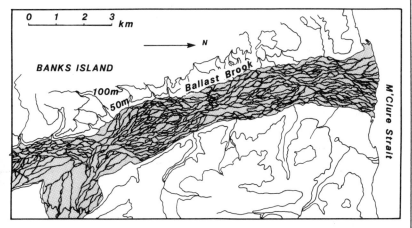

Figure 8.18.
A large braided stream on northwestern Banks Island, western Arctic.

although some vegetated islands, such as those on the South Saskatchewan River in central Saskatchewan, are quite stable. High flow velocities at the junction of two or more channels can create pools up to several metres in depth, but the channels of braided streams are otherwise fairly wide and shallow. This increases bed roughness, the turbulence of the flow, and the ability of the stream to transport sediment.

It has been suggested that braided streams tend to have steeper gradients than meandering streams carrying the same amount of water. A stream could therefore change from meandering to braided if there were an increase in discharge without any change in gradient, or if the gradient increased with no change in discharge. The amount and size of the sediment load in relation to the available discharge, however, are also important factors in determining channel pattern. Braided streams are usually found where stream discharge is highly variable, stream banks are easily eroded, and there is an abundant supply of coarse bedload sediment. They are especially common in Canada on glacial outwash deposits in the Cordillera and in the Arctic, although they are also found in many other areas.

A common type of stream pattern in the Cordillera is transitional between the braided and meandering states. Wandering (laterally unstable) gravel-bed streams consist of stable reaches with single channels alternating with unstable, sometimes braided, reaches with multiple channels. These patterns develop as the supply of coarse, generally glacial, sediment is gradually exhausted (Desloges and Church 1989).

Anastomosed

Anastomosed streams consist of fairly deep and narrow interconnected sand or gravel channels, separated by prominent natural levees, wetlands, and ponds. The channels are of low sinuosity and gradient, and their banks are partially stabilized by the roots of vegetation. Although anastomosed streams are not common, several examples have been recognized in Canada, including tributaries of the North Saskatchewan River in Banff National Park, the upper Columbia River in British Columbia, and the lower portion of the Saskatchewan River straddling the Saskatchewan-Manitoba border at The Pas (Fig. 8.19) (D.G. Smith 1973, 1983, Smith and Smith 1980). Their development requires rapid vertical accretion, which has been about 60 cm per century in the last 2,500 years in the upper Columbia River system, and about 15 cm per century in the last 2,270 years in the lower Saskatchewan River system.

Figure 8.19.
The anastomosed Saskatchewan River in eastern Saskatchewan near the Manitoba border. Areas between stream channels consist largely of wooded swamp.

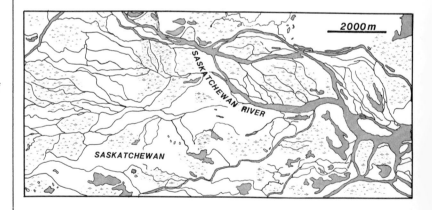

Erosional and Depositional Features

A valley is essentially the product of the stream that flows in it, but as stream downcutting proceeds, many other processes, including weathering and mass movement, play an important role in widening the valley. Such factors as climate (which determines stream discharge and vegetational cover), geology (including rock type and structure), and the available relief determine whether valleys are narrow and V-shaped, or wider, steep-sided, and more U-shaped.

Floodplains

Streams build floodplains by depositing alluvium on their valley floors. A floodplain is a temporary storage area for sediment eroded from the drainage basin. Lateral accretion is accomplished by the continuous growth of coarse-grained point bars on the inside bank of migrating stream meanders, and by the formation of new point bars when the stream changes its course during or following floods. Lateral accretion can also be accomplished by the deposition of fine-grained deposits on the outer concave or cut banks of stream meanders, on the upstream side of the meander axes (Fig. 8.16) (Hickin 1986). These concave bank benches are rare on freely meandering streams, but they are an important component of floodplains where confined meanders with tight bends migrate fairly rapidly downvalley. They have been described on the Squamish, Muskwa, and Fort Nelson Rivers of British Columbia, and are common elsewhere in Canada where underfit streams occupy large meltwater channels; it has been estimated that about 30 per cent of the floodplain on a section of the Fort Nelson River is composed of concave bank bench deposits.

Vertical accretion occurs when streams overflow their banks during floods or, as on some northern Canadian rivers, when they are blocked by ice jams during spring break-up. Coarser sediment tends to be deposited near the stream channel, where it may build up low, ridge-like levees, while the finer material is carried further from the channel and laid down as backswamp deposits. Other sediment is deposited on the floodplain and in the stream channel as the flood waters recede. Although lateral accretion seems to be dominant on most floodplains today, vertical accretion may have been much more important in the past, before the floodplains attained their present elevations. Most of the terraced floodplain sediments in the Porcupine Hills of southwestern Alberta, for example, were deposited by overbank deposition, rather than by lateral accretion (Stene 1980).

Braided streams build floodplains in a similar manner to meandering streams, although the erosion of channel banks and the deposition of bars are not restricted to one particular side of the channel. A continuous floodplain of a braided stream develops through the coalescence of abandoned islands and channels. The braided Alexandra-North Saskatchewan Rivers flow across outwash deposits in Banff National Park. They carry meltwater from the Columbia Icefield, and have been steadily building up their floors for almost 2,500 years (Smith 1973). Deposition on the bed of the channel is the primary process in the braided outwash channels, but overbank deposition of silty sediment dominates in the vegetated backwater areas, where the anastomosed channels are stabilized by the roots of vegetation.

Figure 8.20.
Types of stream terrace.

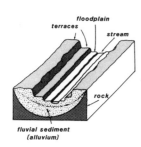

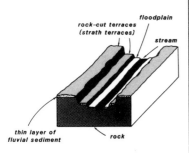

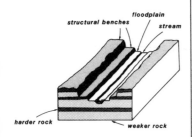

Stream Terraces

Stream terraces are fairly flat bedrock or alluvial benches perched on the sides of valleys (Fig. 8.20). They are the remains of the former valley bottoms or floodplains of streams that once flowed at higher levels than they do today; they are therefore no longer flooded by the streams. Terraces are paired or matched when they are at the same elevation on either side of a valley, and unpaired or unmatched when they are at different elevations. There are several terraces near Lillooet in the middle Fraser Valley of British Columbia. The wider upper terraces (T1, L1, T2 and L2) are unpaired, but the narrower lower terraces (T3-L3 and in some places T4-L4) are paired. The highest terrace (T1) may have been an outwash train formed during deglaciation. Terrace formation began about 7,000 years ago, possibly with the onset of cooler, wetter conditions and an increase in the discharge of the Fraser River (Fig. 8.21) (Ryder and Church 1986).

Terraces are formed when a change in the environment causes a stream to cut down into its floor, or into the material it had previously deposited on the floor. This could be the result of tectonic uplift or tilting, changes in sea level, or changes in climate, which affect discharge and sediment load. The terraces on the sides of spillways, for example, formed when rapid drainage of glacial lakes caused catastrophic flooding and deep incision of the channel floor. Stream terraces are virtually ubiquitous in Canada. The Oldman River Valley has ten sets of terraces in the Rockies of Alberta, and the lower Crowsnest Valley has nine or ten. The North Saskatchewan River at Edmonton rapidly incised its valley in the early **Holocene**, producing a suite of four terraces.

Valleys in glaciated areas were usually infilled with outwash train deposits (Chapter 6), because more sediment was supplied by the ice than the meltwater streams could carry away. When the ice disappeared, changes in stream discharge and sediment resulted in the widespread development of terraces. Nevertheless, not all terraces along the sides of glaciated valleys were formed in this way. The gravel terraces of the Bow River in southern

Figure 8.21.
The Fraser stream terraces at Lillooet, British Columbia (Ryder and Church 1986).

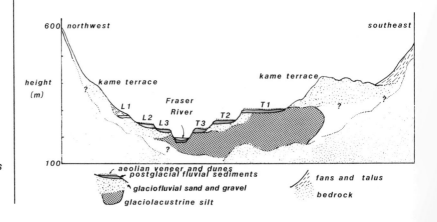

Alberta can be traced for more than 100 km from the Rocky Mountain Foothills to beyond Calgary. Workers now believe that these terraces were not cut into **glaciofluvial** outwash train deposits, but into postglacial sediments supplied to the tributaries of the Bow by debris torrents (Chapter 3) and other agents of mass movement, mainly between about 11,500 and 10,000 years ago (Jackson et al. 1982).

Underfit Streams

Underfit streams appear to be too small for the valleys they flow in. Some types do not meander, whereas others have meanders that are smaller than the meanders of their valleys (Fig. 8.22). The wavelength and radius of curvature of stream meanders are partly determined by the width of the channel and the discharge of the stream when the channel is full. Stream meanders that are smaller than valley meanders may therefore result from a decrease in the amount of water flowing in the stream. Valley meanders in some areas could have been cut when the climate was wetter than today, but in Canada and other glaciated regions they may be attributed to the meltwater running from the shrinking ice sheets of the late **Wisconsin** (Klassen 1972). The smaller stream meanders of today would then reflect an adjustment by the streams to lower amounts of runoff.

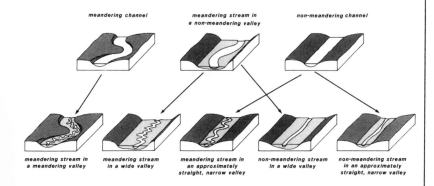

Figure 8.22.
Some types of underfit stream (modified from Dury 1964).

Deglaciation in Canada caused rapid hydrological changes. Large valleys that once carried far more water than they do today are now occupied by small, local streams. Several streams may flow in different segments of a large spillway. The Milk River in southern Alberta, for example, flows through a valley that today appears to be far too deep and wide in relation to the present size of the river. This large valley was cut when the Milk River was receiving nearly all the runoff in southwestern Alberta, because drainage to the northeast was blocked by ice. Retreat of the Laurentide ice sheet eventually allowed the modern South Saskatchewan River to divert most of the runoff into Hudson Bay, leaving the Milk River as an excellent example of an underfit stream.

Deltas

Deltas are formed where a stream drops its load as it enters a body of standing water, such as the sea or a lake (Chapter 9). They generally fail to develop in areas exposed to strong waves, currents, or tides, or if the stream carries too little sediment. The shape of deltas varies according to the stream regime, coastal processes, climate, and the structural stability of the coast. They can be triangular, as described by the Greek letter delta (Δ), fan- or arcuate-shaped, or composed of many branching channels in the form of a bird's foot. Other deltas, including that of the Mackenzie, have been built in the valleys or estuaries of drowned streams.

Fans

Fans are deposited on land, usually where a stream leaves the confines of a mountain valley and flows out over a flat plain. They are the result of the decline in a stream's ability to carry sediment because of the sudden increase in channel width that is now possible, and the consequent reduction in depth of flow and velocity. Fans tend to be cone-shaped, with their apex at the point where the stream leaves the mountain front (Kostaschuk et al. 1987). The gradient and the size of the sediment on a fan tend to decrease radially outwards from its apex. Fans generally consist of sand and gravel, but fine-grained silt is the main component of at least one fan in the northern Yukon. This may partly reflect the importance of frost disintegration in this area.

Paraglacial origins • The term 'paraglacial' has been used to describe the transition between glacial and nonglacial conditions (Ryder 1971a, Church and Ryder 1972). Paraglacial processes include the rapid wastage of unstable glacial and periglacial material from recently deglaciated slopes (Johnson 1984a), the transportation of these sediments in streams (Church and Slaymaker 1989), and their deposition in fans. Paraglacial deposits can be the fairly recent products of alpine glaciers. The Bella Coola River in British Columbia, for example, has become more stable as the supply of sediment from the erosion of eighteenth- and nineteenth-century (**neoglacial**) moraines has become exhausted. Fans in the Cordillera were often formed fairly quickly by stream flow or debris torrents in the transitional or paraglacial period between glacial and nonglacial conditions. Deposition of coarse sediment began as soon as an area became ice-free, often where a tributary from a hanging valley entered the main valley. The fans in the Bow Valley in Banff National Park were almost completed about 6,000 years ago, and the major period of deposition in south-central British Columbia seems to have ended about 6,600 years ago (Ryder 1971b). These fans are essentially inactive today, and are being dissected by stream action. Deposition has probably stopped because of reductions in the amount of available glacial sediment and, with the retreat and disappearance of the glaciers, the amount of water carried by the streams during floods.

Debris-flow-dominated and fluvially dominated fans • Fans in many areas are still active. Two types can be distinguished in the Foothills of the Rocky Mountains (Kostaschuk et al. 1986). Debris flows (Chapter 3) are the main depositional mechanisms on fans at the mouths of basins on the steep northeasterly-facing **scarp slopes**, while fluvial processes are most important on the larger fans associated with the more gently sloping and larger basins on the southwesterly-facing **dip slopes**. The surface of the debris-flow fans is characterized by lobes, levees, and concentrations of boulders, the deposits are poorly sorted, angular, and unstratified, and the fans are small and steep. Fluvial fans have gravel-bed streams with stratified, rounded stones. The fluvial fans are larger than the debris-flow fans because their larger basins supply more sediment. They also have lower gradients than debris-flow fans, reflecting the greater ability of streams to transport sediment over gentle slopes.

The Longitudinal Profile of Streams

Although streams generally have smoothly concave longitudinal profiles with slope declining progressively downstream, many Canadian streams still have typically irregular glacial profiles consisting of a series of convex steps and concave basins (Fig. 8.23).

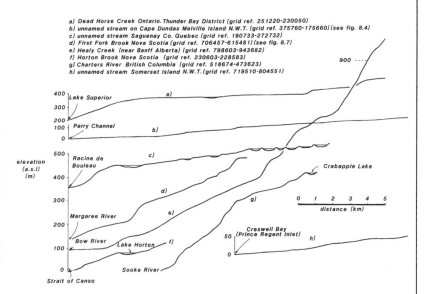

Figure 8.23.
Examples of Canadian stream profiles.

Waterfalls

Waterfalls and rapids occur along the steeper sections of stream profiles, particularly where there are outcrops of more resistant rock (Fig. 8.24). Streams tend to remove irregularities, producing smooth, concave-upward profiles with a progressive downstream reduction in gradient. Erosion is

Figure 8.24.
Some types of waterfall and rapid (modified from Selby 1985).

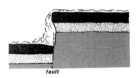

fault

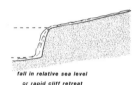

fall in relative sea level
or rapid cliff retreat

■ hard rock

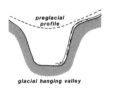

preglacial
profile

glacial hanging valley

rapid in some cases. On the Pembina River in northwestern Canada, for example, a waterfall retreated upstream by more than 300 m in nine years. Waterfalls are therefore transitory features, which may become rapids as a result of lowering and retreat, before eventually disappearing.

The Niagara River partly exhumed a mid-Wisconsin gorge, which was filled with late Wisconsin till and **lacustrine** deposits (Fig. 8.25). Other buried drainage routes may have influenced the development of the modern Niagara River, but the only proven re-occupation is of the St Davids Gorge between the Whirlpool and the head of the Whirlpool Rapids. The rate of retreat of the Falls has varied with the discharge of the Niagara River, reflecting changes in the spillways used by the developing Great Lakes system (Chapter 5). Natural retreat of the approximately 50 m high Horseshoe Falls is about 1 to 2 m per year, but the diversion of water to generate electricity and other modifications made to stream flow have probably reduced the rate of retreat to less than 0.03 m yr^{-1} today.

The Steady State

Streams try to maintain a balance between their capacity and competence and the amount and type of sediment they carry. The slope of a graded stream is just sufficient, measured over a period of time, for it to move its sediment load; there is therefore no net erosion or deposition in the channel. As one moves down a stream course, grain size decreases and the discharge increases as tributaries feed their flow into the main channel. It has therefore been argued that stream profiles are concave because the greater flow makes it increasingly possible to transport the fine sediment load downstream over lower gradients. Nevertheless, it is unlikely that any single factor can explain the general concavity of stream profiles.

Concave profiles may represent the graded or equilibrium form of a stream, although some graded streams can have quite irregular profiles. This steady state is achieved and maintained through the interaction of channel characteristics, including width and depth, gradient, bed roughness, and channel pattern. Changes in climate or vegetational cover, which affect stream discharge and sediment size and amount, cause compensatory changes in the channel characteristics. These changes serve to accommodate the effect of the original changes and to restore the steady-state condition. The meandering Red River near Winnipeg, for example, is becoming straighter. This may be because of isostatic rebound to the north, which is reducing slope gradient, a coarsening of the bed-load material, which increases the roughness of the bed, or artificial control of water levels at a dam. Similarly, channel straightening and widening in the United States have been attributed to increased peak discharges and sediment loads resulting from nineteenth-century settlement. It is likely that man's effect on the streams of southern Canada, through such activities as agriculture, forestry, urbanization and water management, is of similar magnitude.

There has not been enough time since the end of glaciation for many

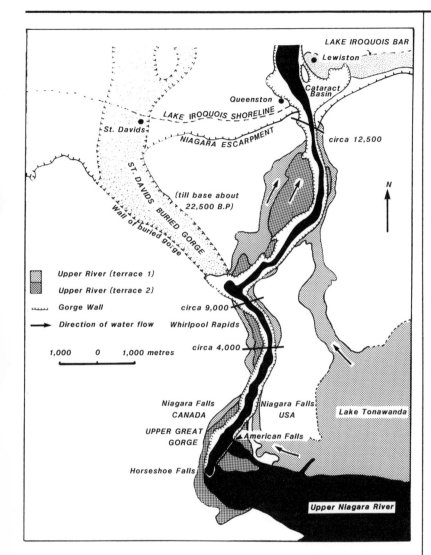

Figure 8.25.
*Niagara Falls (Tinkler
1986). The approximate
position of the retreating
falls is shown at various
times.*

Canadian streams to attain an equilibrium profile (Fig. 8.23) (Ponton 1972). The relationship between channel width, channel depth, flow velocity, and discharge on Nigel Creek, for instance, a hanging tributary of the North Saskatchewan River in the Rocky Mountains, suggests that these factors are in quasi-equilibrium with modern conditions. But the absence of any systematic decrease in channel gradient downstream and the presence of major changes in slope along its course show that this typical mountain stream is still in the process of adjusting its gradient to geological and glacial influences (Knighton 1976).

One may distinguish three kinds of stream:

(a) those in equilibrium over their full length. There are very few streams of this type in Canada.

(b) those locally in equilibrium over sections extending between boundaries that are usually geologically or glacially induced breaks in slope. These streams too are of three types:

(i) those that are simply equilibrated.

(ii) those that first became equilibrated under paraglacial conditions, then were re-equilibrated as vegetation and soil developed, or as a result of natural catastrophic events, including landslides. These streams are common to dominant in mountain valleys, and in the north.

(iii) those of type (i) or (ii) that are re-equilibrating to the impact of European colonization, including deforestation, agriculture, channelization, diversion, removal of groundwater, etc. These are common in the populated areas of southern Canada.

(c) those that are totally unequilibrated. These are common to dominant in mountain heads.

Coastal Processes and Landforms

9

The coastline of Canada extends through about 40° of latitude, winding for almost a quarter of a million kilometres along the shores of three oceans and four Great Lakes. Canada has the world's longest coastline, as well as more estuaries and deltas than any other country, and more fiords than all other countries combined. There are many different types of coastal environment in Canada, representing an enormous range of geologic, climatic, and oceanographic conditions (Owens and Bowen 1977, McCann 1980, Bird and Schwartz 1985). Waves, for example, are frequent and intense on the Pacific and, to a somewhat lesser extent, Atlantic coasts, but they are generally weak and infrequent on the ice-infested coasts of the Arctic. Similarly, while the tides in the Bay of Fundy are the highest in the world, in the Great Lakes they are almost non-existent. Furthermore, while many coastal processes operate in much the same way in Canada as elsewhere, factors such as **isostatic** recovery, coastal permafrost, ground ice, and sea ice assume special importance in the Canadian context.

Changes in Sea Level

The diversity of the Canadian coastline is enhanced by the presence of deltas, gravel ridges, beaches, wave-cut terraces, and other features of abandoned shorelines, formed when sea level was higher than today relative to the level of the land. There are many possible reasons for variations in **relative sea level**, including changes in the volume of the ocean basins, in the level of the land, and in the position of the Earth's poles, the tilt of its rotational axis, and its rate of rotation. Some of the most rapid major changes, however, have been the result of the alternate growth and decay of ice sheets. The growth of ice sheets reduces the amount of water in the oceans, as it is converted into glacial ice by way of the hydrological cycle (Fig. 8.1). The amount of water in the oceans then increases at the end of **glacial** stages, when meltwater returns to the ocean basins. One would therefore normally expect sea level to be lower during glacial than **inter-glacial** stages.

Eustatic sea level appears to have fallen in the latter part of the Tertiary period, possibly as a result of the growth of Antarctic ice. Large fluctuations in sea level in the Pleistocene, however, and particularly during the last 700,000 years, can be largely attributed to the periodic growth and decay of ice sheets in the northern hemisphere (Fig. 4.6a). Glacial sea levels were between about 80 and 130 m below today's level in many parts of the world, and apparently 200 m or more in some others. High interglacial sea levels were similar to today's on several occasions, but they have been higher only once in the last 700,000 years, during the last interglacial stage about 125,000 years ago. Sea level at that time was from 3 to 10 m higher than today, reflecting higher temperatures and therefore the presence of even less ice on Earth than at the present time. Canadian evidence for this warm period includes the Don Beds of the Toronto area (Chapter 5), and a widespread glaciated and, in places, till-covered rock platform between about 4 and 6 m above present sea level. This coastal platform has been traced along the coast of Newfoundland, Nova Scotia, and the Magdalen Islands, and along the banks of the lower St Lawrence. Wave-cut benches of similar elevation have been recognized in eastern Baffin Island.

Although the amount of water in the oceans gradually increased as the ice melted at the end of the last glacial stage, changes in relative sea level in glaciated regions were also affected by glacially induced changes in the level of the land. **Glacio-isostasy** involves the depression of the land under the weight of the ice during glacial stages, and uplift during warmer periods of ice decay. The amount and direction of postglacial changes in relative sea level in Canada therefore depended upon local ice thickness and the history and timing of deglaciation. The amount of uplift that has taken place since the disappearance of the ice generally increases from the margins of the former ice sheet to the interior, where the ice was thickest. Relative sea level fell in areas that were depressed under thick ice, because the land rose faster than the sea. In more marginal areas, either sea level rose faster than the isostatic rebound of the land, or the land was able to attain its preglacial elevation quickly; this resulted in a rise in relative sea level. Glacially depressed regions that could not rebound fast enough to counter the global rise in sea level were therefore flooded (Fig. 5.8).

In the Atlantic provinces, areas well beyond the ice margin on the outer shelf experienced only a short period of rebound, followed by continuous submergence as sea level rose. Uplift in areas that were much nearer, or even at, the ice margins, including Prince Edward Island, southern New Brunswick, and Newfoundland, was greater than the rise in sea level until about 7,000 years ago. Emergence of the land was then gradually replaced by submergence, as the waning uplift became slower than the rise in sea level. Relative sea level began to rise only quite recently in areas such as western Newfoundland, which were under thicker ice than to the south, and therefore experienced a more prolonged period of uplift. Even further north, in the northern Gulf of St Lawrence, the land has still not completed the uplift or emergence phase. In general, therefore, net subsidence or rise in relative sea level in the southern part of the Atlantic provinces has drowned coastal forests, peat beds, and archaeological sites, while thicker ice and greater depression of the land caused net emergence or fall in relative sea level north of a hinge line in the northern Gulf of St Lawrence (Fig. 9.1).

Uplift of the land in the eastern part of Hudson Bay has resulted in the formation of nearly two hundred raised beaches. They extend up to 200 m or more above present sea level, and represent a period ranging from about 8,500 years ago to the present time. Iron rings used to tie up trading ships in the Churchill area in the early 1700s are now stranded high up on the walls of a dry inlet. Relative sea level in the **Holocene** in eastern Baffin Island reached its highest point about 8,000 years ago, when local ice advance may have depressed the land or brought about a period of crustal stability. There is no evidence of this rise in sea level on southern Ellesmere Island, but a **transgression** may have occurred here about 5,000 years ago. A beach formed at that time in the Arctic Archipelago is now more than 25 m above present sea level.

The amount of isostatic depression in British Columbia also varied with the distance from the former ice centres. The highest point reached by the sea in postglacial times was at least 200 m above its present level near Vancouver and Kitimat on the mainland coast, but only 75 m at Victoria, which was further from the main ice centres. Near the ice margins on the Queen Charlotte Islands, the rise in sea level became greater than the rate of uplift of the land in the late Pleistocene-early Holocene, but this did not occur until the mid-Holocene on the mainland and on eastern Vancouver Island, where there had been greater depression of the land.

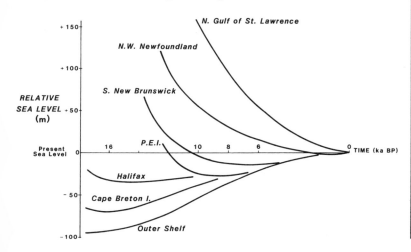

Figure 9.1.
Changes in relative sea level in eastern Canada (Grant 1980). Differences between areas reflect differences in the rate and amount of isostatic recovery according to the local ice thickness and the timing of deglaciation. (ka BP: thousands of years before present.)

Waves

Waves are produced when the wind disturbs the surface of the sea. In deep water, their size is determined by the speed of the wind, the time that it blows in one direction, and the fetch, or distance, of open water over which it blows. Winds generate a confused mixture of waves of different sizes, but sorting takes place as the longer, faster waves emerge from the storm area before the shorter, slower waves. The Pacific coast of Canada and the southern Atlantic coast to about midway up the coast of Labrador can be considered to be storm-wave environments. In these areas, short, high waves are frequently generated by local gale-force winds in winter, and less often in summer. Because westerly winds are dominant in the mid-latitudes of the northern hemisphere, waves on the Pacific coast are probably somewhat higher and more numerous than on the Atlantic coast.

Each water particle in waves in deep water rotates in an essentially closed, circular orbit. The diameter of the orbit of a particle on the water surface is equal to the height of the wave. The size of the orbits of particles beneath the surface diminishes rapidly with depth, however, and movement is negligible at a depth of about half the wavelength of the wave (Fig. 9.2).

When a wave enters the increasingly shallow water near a coast, its orbital motion eventually extends all the way to the bottom. This results in the gradual transformation of the wave, involving a progressive decrease in wave velocity (celerity) and therefore wavelength, and, following an initial decrease, an increase in wave height; **wave period**, however, remains constant.

Wave crests bend or refract when a portion moving through shallower

Figure 9.2.
Orbits of particle movement in a wave, showing the gradual decrease in radius with depth (Pethick 1984). λ is the wavelength and H is wave height.

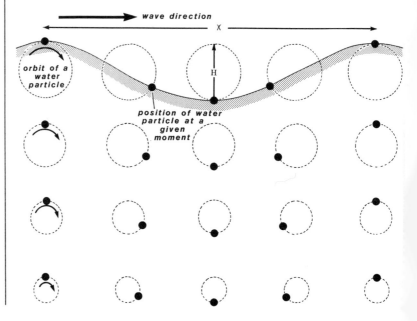

water slows down more rapidly than a portion in deeper water. Waves therefore tend to assume the shape of the coast or, more specifically, the shape of the submarine contours. Refraction causes wave energy to be concentrated on headlands and dissipated in bays. Waves on headlands are therefore higher than in bays (Fig. 9.3). Because of the difference in wavelength and in the depth at which they begin to feel the bottom, the refraction of long, far-travelled swell waves is greater than that of short, locally generated storm waves.

The shape of the orbits also changes as waves enter shallow water. They become more elliptical, and the forward movement of water under the high, narrow wave crests becomes increasingly greater, though of shorter duration, than the seaward movement under the long, flat troughs. Wave transformation in shallow water ultimately causes it to break, in a way that is determined by the wave steepness and the slope of the bottom. Wave transformation in shallow water and the characteristics of the breaking wave have important implications for the generation of currents and the movement of sediment (Fig. 9.4).

Although water particles continue to rotate in orbits as waves enter shallow water, there is now some slight movement, or mass transport, of

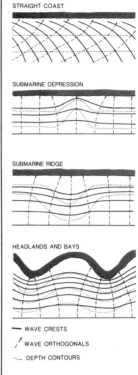

Figure 9.3.
Wave refraction diagrams. The wave crests represent the waves, while the orthogonals show their direction of travel.

STRAIGHT COAST

SUBMARINE DEPRESSION

SUBMARINE RIDGE

HEADLANDS AND BAYS

— WAVE CRESTS

/ WAVE ORTHOGONALS

--- DEPTH CONTOURS

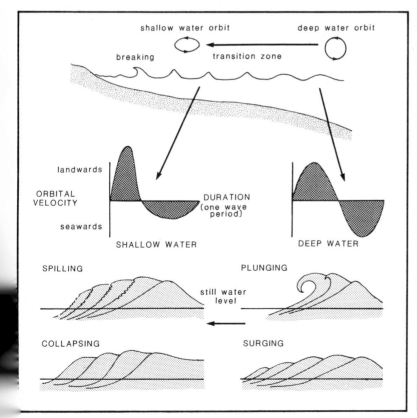

Figure 9.4.
The transformation of waves in shallow water, and types of breaking waves.

the water towards the coast. The water that is built up at the shore may flow seawards through concentrated return flows or rip currents, particularly during periods of beach **erosion** or **deposition**. Rip currents flow seawards where the breaking waves are slightly lower than elsewhere. These lower points can be produced by wave refraction, but in most cases they appear to be associated with a kind of standing wave known as **edge waves** (Fig. 9.5) (Huntley 1980).

Currents flowing alongshore or parallel to the coast in the **surf zone** can be generated by waves breaking at a slightly oblique angle to the coast, or by large-scale variations in the height of the breaking waves along a coast. Longshore currents are strongest just landwards of the **breakpoint**, and they are absent at the beach and at a short distance seawards of the breakpoint. On many coasts, the superimposition of currents that are parallel and perpendicular to the beach forms a cell circulation pattern, consisting of slow mass transport, and longshore currents feeding strong, narrow rip currents (Fig. 9.5). In the southern Gulf of St Lawrence these circulation cells have usually developed by late June, and they then persist for the rest of the year (McCann and Bryant 1972).

Figure 9.5.
Edge waves, cell circulation, and rip currents. The rip currents are shown operating in areas where the edge waves produce slightly lower crests on the incoming waves.

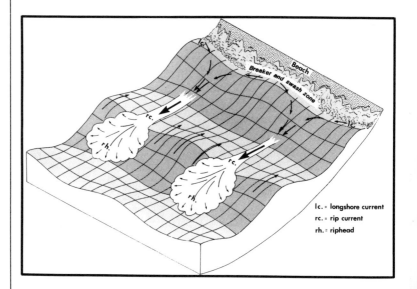

lc. = longshore current
rc. = rip current
rh. = riphead

Tides

Tides are very long waves generated on Earth by the gravitational attraction of the Moon and, to a lesser extent, the Sun. One water bulge forms on the side of the Earth directly under the Moon, and another on the opposite side. Because of these two bulges, the twenty-four hour rotation of the Earth around its axis, and the movement of the Moon around the Earth, a body of water would normally experience two high and two low tides every

24 hours and 50.47 minutes. Semi-diurnal tidal regimes therefore have two, albeit unequal, tidal cycles every 25 hours, but some areas have diurnal regimes with only one tidal cycle during this period. Other areas experience a mixture of semi-diurnal and diurnal tidal cycles.

Tides are also influenced by the gravitational attraction of the Sun, although its tidal generating force is only about half that of the Moon. Tides vary according to the relative positions of these bodies as the Moon travels around the Earth in a 29-day orbit. The maximum tidal-generating force, producing the greatest range of high and low tides, occurs twice a month, when the Sun, Moon, and Earth are roughly aligned. These maximum or spring tides are matched by two neap tidal periods, when the opposition of the gravitational attractions of the Moon and Sun produces tides with a minimum tidal range (Fig. 9.6).

Tidal range, which is the difference in the elevation of the high and low tidal levels, is usually less than 2.5 m in shallow coastal waters. It is much greater in more restricted waters, however, as in the bays and inlets of northern British Columbia, southern Baffin Island and Ungava Bay, and particularly in the Bay of Fundy, where the maximum range is more than 15m. These high or macrotidal ranges contrast with negligible or microtidal ranges in the Great Lakes, and low, micro- to mesotidal ranges in the western Arctic, eastern Baffin Island, eastern Hudson Bay, and the Gulf of St Lawrence (Fig. 9.7) (Dohler 1966).

Figure 9.6.
The tidal forces in a lunar cycle (Pethick 1984).

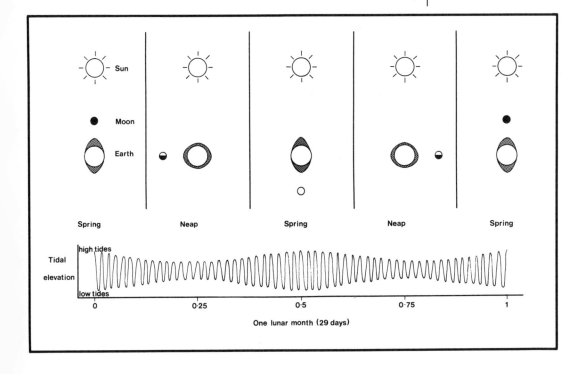

Sun

Moon

Earth

Spring   Neap   Spring   Neap   Spring

Tidal elevation   high tides

low tides

0   0·25   0·5   0·75   1

One lunar month (29 days)

Figure 9.7.
*Tidal range in Canada. The
map was drawn from data
published in the* Canadian
Tide and Current Tables
*(Canadian Hydrographic
Service, Ottawa), supple-
mented with information in
Dohler 1966.*

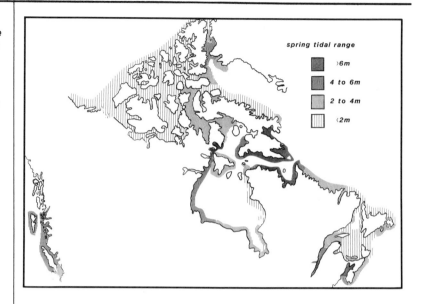

Tides exert a strong influence on wave processes, and consequently on beach sedimentation and erosion. Where there are high tidal ranges, wave energy is less concentrated at any particular elevation, and the water level changes faster and by greater amounts than in areas with small ranges; tidal currents, however, are strongest where the tidal range is high. Very strong tidal currents can be generated in macrotidal estuaries and bays. When the tide is rising, it rises faster in the sea, a large body of water, than in coastal inlets. The difference in the level of the water produces strong inward-flowing currents, which move large amounts of sediment into the inlets. When the tide is falling, it falls faster in the sea than in the inlets, producing seaward-flowing currents. Tidal currents are the dominant coastal mechanisms in the Bay of Fundy, and in other areas with very large tidal ranges, and they may eventually be harnessed on a large scale to produce electricity.

Although many factors influence the development and distribution of coastal features, some seem to be characteristic of areas with small tidal ranges, whereas others are associated with high ranges. River deltas, barrier islands, spits, and sub-horizontal shore platforms in Canada and elsewhere, for example, tend to occur where the tidal range is small. Salt marshes, mudflats, and sloping shore platforms, on the other hand, are generally best developed where the tidal range is large.

Ice-dominated Coasts

For convenience, one may distinguish ice-dominated, low-energy environments from wave-dominated, storm-wave environments, but it must be emphasized that all of Canada's coasts, with the exception of British Columbia and the northern portions of the High Arctic Islands, experience a combination of wave and ice action.

About 90 per cent of Canada's coastline is affected by ice. Most of this ice is seasonal, but there is **perennial** polar **pack ice** along the northern coast of the High Arctic Islands (Fig. 9.8 A). Coastal features formed by ice are most prominent in Arctic and subarctic regions. They are also found in southern Canada, although they are generally less well developed and more quickly obliterated by waves and currents in the open-water season. In the Bay of Fundy and eastern New Brunswick, for example, it has been found that there is no visible evidence of ice action on the beach after only a few weeks of ice breakup (Knight and Dalrymple 1976, Owens 1976).

The presence of ice off a coast prevents the formation of waves, or limits the fetch distance, whereas ice in the coastal zone and on the beach absorbs wave energy and reduces the effectiveness of any wave action that does occur. Ice in the channels between the islands of the northern and western Arctic still inhibits wave action as late in the year as mid-August. Sea ice, together with fluvial and other land-based processes, therefore dominates in the coastal zone of the northwestern Arctic. Coastal sediments are derived from the local rocks in this area, because the waves are generally unable to move material for any great distance along the coast. Waves become more important to the east in the Arctic Archipelago, as the length of the ice-free season increases (Taylor and McCann 1983). Although there are well-developed depositional features in the east-central Arctic, however, waves become much more effective further east, with an increase in tidal range and in the duration and extent of open water. Nevertheless, beaches throughout the Arctic tend to be narrow and poorly developed, consisting of coarse sand and cobbles that are poorly sorted and rounded in comparison with those found on beaches in the more vigorous wave environments further south.

Because of the presence of permafrost beneath Arctic beaches, storm waves and ice are usually able to move only sediment lying at depths of

Figure 9.8.
A) Distribution of sea ice cover in the Arctic in mid-winter, and B) in mid-August (Taylor and McCann 1983).

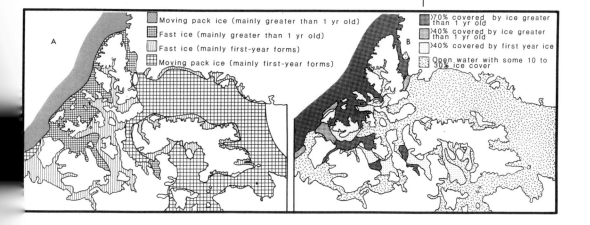

less than 90 cm. Nevertheless, large waves may occasionally remove the unfrozen beach sediments and expose the underlying permafrost, resulting in the thawing and erosion of the ice-bonded sediments; this happens about every two or three years in the eastern Arctic. **Thermal erosion** of sediments containing massive ground-ice wedges or lenses also causes slumping and retreat of coastal bluffs (Chapter 3). Retreat is particularly rapid in ice-rich silts and clays, which do not provide suitable material for the development of protective beaches at the foot of the bluffs. Ice-rich bluffs along sizable portions of the mainland coast of the Beaufort Sea and southwestern Banks Island are retreating at rates of more than 2 m per year, and maximum rates are locally much greater. This erosion is accomplished within the three-month ice-free period.

The 'ice year' can be divided into four periods, characterized by freeze-up, development of a complete ice cover, ice breakup, and the return of open water (Taylor and McCann 1983):

(a) Freeze-up at the shore and the formation of an **icefoot** takes place in autumn, with the freezing of wave **swash**, spray, and **interstitial** water, the stranding of **ice floes**, and the accumulation of ice slush. In tidal areas, the icefoot can extend down to the low tidal level, and on exposed coasts it can also extend to well above the high tidal level. A 'kaimoo' may develop in the upper part of a beach in an area with a small tidal range; these deposits consist of alternating layers of ice and beach sediment produced by the freezing of swash and wind-blown sand.

(b) **Fast ice** completely covers the littoral zone in winter. It fills the channels between the islands of the High Arctic, and can extend more than 100 km off the coast of eastern Baffin Island. The alternate grounding and floating of this ice as it rises and falls with the tide may be an important process where there is a significant tidal range.

(c) The breakup begins with the melting of the snowcover and the enlargement of tidal cracks in the ice. The rapid melting of fractured, dirty ice in the **intertidal** zone usually results in the development of a wide break between the more persistent icefoot and the offshore sea ice.

(d) The beach sediments are then reworked by waves during the period of fairly open water, which decreases northwards and westwards from the eastern Arctic (Fig. 9.8 B).

In addition to its indirect effect in limiting wave action, ice has a direct effect on beach development. Nevertheless, it should be noted that although the pushing and melting of ice produce conspicuous features in cold coastal environments, they affect only a small proportion of the beach material, and do not greatly alter the overall shape of the beach.

Ice rafting involves the transportation and deposition of sediments by ice floes, which are driven by waves, winds, and currents. Rafting can be accomplished by the refloating of grounded blocks of ice and the movement of sediments that are frozen onto their bases. The process moves large amounts of muddy sediment in the tidal environments of eastern Canada, but it is less effective where the sediments consist of coarser sand or gravel. Ice can also move boulders for short distances across tidal flats, although it is not clear to what degree it does so by rafting, as opposed to sliding, pushing, or rolling them along. Wide tidal flats are commonly littered with boulders and cobbles lying on, or buried in, sand or other finer ice-rafted sediments. Boulders are randomly distributed across broad intertidal zones in flats, pavements, and fields, but they can also be found in a nearly continuous, partly submerged row or barricade near the low tidal level. Although boulder barricades may be formed in different ways in different areas, most workers believe that the boulders are deposited or pushed into place by grounded, wind-driven slabs of ice; the presence of an icefoot or shore-fast ice prevents the boulders from being moved further landwards. Boulder barricades are particularly well developed in Labrador, but they also occur around Hudson and Ungava Bays, in southern and eastern Baffin Island, and even as far south as the St Lawrence Estuary (Rosen 1979, McCann et al. 1981).

A variety of ridges and scour marks is produced by pack ice and large floes along most shores in high latitudes, especially on headlands, at the mouths of rivers, and on exposed tidal flats. Ice driven onshore by the wind pushes up ridges and mounds in the intertidal and **supratidal** zones, often producing a corresponding depression in the lower parts of the beach. The best development of these ridges is in areas where there is moving pack ice for a large part of the year, particularly along the western side of the Arctic Archipelago and in the Parry Channel. They are much less common between the islands in the High Arctic, where winter ice is fast and breaks up quickly in summer. Rocky coasts, higher tidal ranges, and the occurrence of boulder barricades inhibit the formation of ice-pushed ridges in the eastern Arctic.

Ice scour and other ice-related processes in the offshore zone produce an irregular sea bed with pits, scour marks, ridges, and shoals. Scours can be made by floating ice, the bulldozing of grounded ice, or boulders pushed by the ice. The scouring of salt marshes in the St Lawrence Estuary produces barren depressions that may develop into marsh ponds or pans. Similar features are formed when ice tears away large blocks of peat on the tidal marshes of James and Hudson Bays. They include long furrows cut by boulders dragged over the marshes by ice floes, and circular depressions, sometimes with a surrounding rim, where ice floes were alternately raised and lowered by waves and the tide. These processes contribute to the development of patterned marshes with a jigsaw-like appearance in some areas and, in others, marshes with long, narrow, straight, and deep channels. Various pits, mounds, and sedimentary structures are also formed on beaches by the melting of the icefoot, partially buried ice boulders and sea ice driven onshore during storms. These features will persist, however, only if the pitting is extensive and occurs above the level of the high tides (Dionne 1978, 1988, Martini 1981a).

Wave-dominated Coasts

While waves can play an important role in the Arctic during occasional ice-free periods, in most of southern Canada they are the most important coastal mechanisms.

Sand Beach Profiles

The character of a beach varies according to such factors as the type of sediment and the size of the incoming waves. Beaches are very dynamic landforms, able to adjust to variations in waves, tides, and other influences. Two profiles represent the extremes of the range of forms that a beach may assume according to the size or power of the waves. Steep beaches reflect much of the incoming wave energy. These reflective systems have well-developed **berms** and rhythmic **beach cusps**. A pronounced step of coarser sediment forms where the incoming waves meet the **backwash** (Fig. 9.9 f). At the other end of the spectrum, dissipative systems absorb or consume much of the incoming wave energy before the waves reach the beach. These beaches have concave-upward nearshore zones and wide, flat surf zones. Their profiles are usually more complex and varied than reflective systems, and there are generally one or more submarine ridges or bars (Fig. 9.9 a). Several intermediate states, distinguished by differences in morphology, currents, and sediment transport, may be identified between the fully dissipative and fully reflective systems (Fig. 9.9 b-e).

The shape of a beach varies in space and time according to changes in wave energy. During a storm, increasing wave power causes beach erosion, formation of a submarine bar, and transformation of the beach from a reflective to a dissipative state. Decreasing wave power, possibly in the period following a storm, causes the bars to migrate shorewards, and the system gradually changes from a dissipative to a reflective or fully **accreted** state (Fig. 9.9 a-f). In the storm-wave environments of southern Canada, winter waves are consistently powerful and sandy beaches are therefore usually in a highly dissipative state, although in summer, more prolonged periods of calmer conditions may allow some transformation into steeper reflective states, especially in bays and other sheltered areas (Bryant 1983).

The term 'ridge and runnel' was originally used in Britain to describe a series of low intertidal ridges and depressions. These undulations develop on gently sloping beaches in areas with a large tidal range but only limited exposure to wave action. The ridges are quite stable, maintaining their shape and relative positions through time. Unfortunately, the term has recently been used in North America to refer to ridges or bars in the **swash zone** of tidal and non-tidal coasts. These ridges are moved landwards by the lower energy waves following a storm (Fig. 9.9 e). Storm-induced 'ridge and runnel' is a common feature of the stormy oceanic and inland coasts of Canada. The traditional or stable type of ridge and runnel, however, has also been recognized in the Maritimes, and up to fourteen ridges

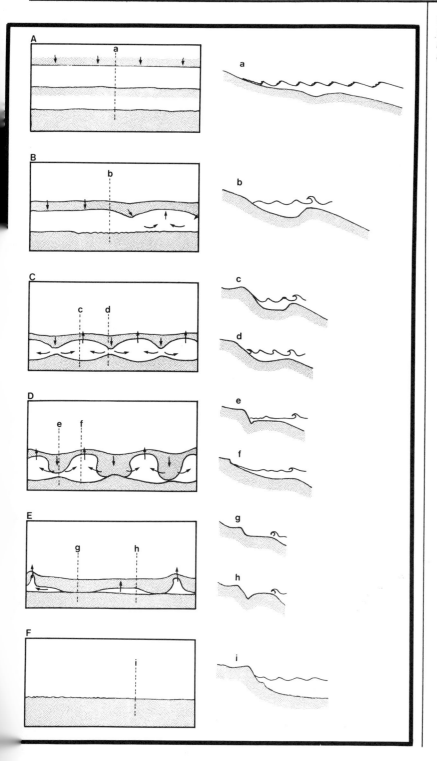

Figure 9.9.
*Plan and profile of six
major beach states (Wright
and Short 1984).*

have been reported between mean water level and the low tidal level in Craig Bay on Vancouver Island (Hale and McCann 1982).

Although submarine bars can be formed by low waves following a storm (Fig. 9.9 b), they can also develop in other ways. Bars can be parallel or perpendicular to the coast, straight or **crescentic** (Huntley 1980), stable or mobile, single or multiple, and intertidal or **subtidal**. There are two stable bar systems in Kouchibouguac Bay in eastern New Brunswick (Greenwood and Davidson-Arnott 1975). The inner system consists of one to three bars, from 0.5 to 1.25 m in height. These bars can be straight, oblique, or crescentic, and they are dissected by channels excavated by rip currents. A more continuous outer bar, 1.5 to 3 m in height, has a rhythmic crescentic appearance with the horns facing shorewards. Freeze-up prevents wave action for several months and destroys or disrupts the submarine profile formed by waves in the nearshore zone. The outer bar probably experiences little modification by ice, however, and the bars and troughs of the near-shore zone are re-established by waves in the spring. The Great Lakes also contain a variety of stable and unstable bars. There are up to nine low submarine ridges parallel to the shoreline in southern Georgian Bay. Unlike other bars that have been studied in the Great Lakes, these do not migrate onshore and are not destroyed by storms (Davidson-Arnott and Pember 1980).

Pebble Beaches

Stony beaches are found on all of Canada's coasts, but they are particularly common in the Arctic. Large amounts of coarse sediment are produced by glacial and periglacial processes in the north, and, at least in southern Canada, by the erosion of bedrock and glacial deposits by waves. Large breaking waves can throw pebbles beyond the limits of the swash, building ridges at the back of the beach to considerable heights above the high tidal level. These storm ridges often develop across the mouths of small streams, diverting the flow of water to the sea. Beach slope generally increases with the grain size of the sediment. Beaches of well-sorted pebbles are therefore much steeper than those where sand, or sand mixed with pebbles, are the main components. At Advocate Harbour in the Bay of Fundy, for instance, the pebble beach has an average gradient of about 6.8°, compared with slopes of between 1.3 and 1.8° on the sand beaches of Nova Scotia (Taylor et al. 1985). The steepness of pebble beaches is the result of the rapid percolation of the swash into the beach, and the corresponding weakness of the backwash. These steep beaches therefore tend to be in reflective states, with poor development of bar and trough topography, although they are often fronted by more gentle, dissipative beaches of sand.

Spits, Barriers, and Other Beach Forms

Beaches are usually attached to the land along their entire length, but some types of beach are at least partly detached. Such features usually develop where there are sudden changes in the direction of the coast, obstructions to the longshore flow of material, or sheltered (wave shadow) areas, such as behind islands. Submerged glacial moraines running across bays and estuaries may also have provided the foundation for some features in Canada. The shape of these beach forms is affected by changes in relative sea level, erosional retreat of the coast to which they are attached, changes in the strength and direction of the waves, and variations in the amount and source of the sediment being supplied.

Classification • Beaches can be classified in a number of ways, using a variety of criteria and terms. The simplest approach is based upon how they are attached to the land, although this can result in use of the same term for features that were formed in different ways and in different situations. Omitting normal beaches, three major groups can be identified (Table 9.1):

Table 9.1 Beach types*

a) *Beaches attached to the land at one end*
   1. Length greater than width
     (a) Continuation of original coast, or parallel to the coast[1] (spits)
     (b) Extending out from coast at high angles[2] (arrows), or
        extending out from the lee side of an island (comet-tail spits).
   2. Length less than width (forelands)

b) *Beaches attached to the land at two ends*
   1. Looped forms extending out from the coast
     (a) extending from lee side of an island (looped barriers)
     (b) a spit curving back onto the land[3] or two spits or tombolos
        joining up[4] (cuspate barriers).
   2. Connecting islands with islands or islands with the mainland
     (tombolos)
     (a) single form (tombolos)
     (b) single beach looped at one end (Y-tombolos)
     (c) two beaches (double tombolos)
   3. Closing off a bay or estuary (barrier beaches)
     (a) at the mouth (front) of a bay (baymouth barriers)
     (b) between the head and mouth of a bay (midway barriers)
     (c) at the head (back) of a bay (bayhead barriers)

c) *Forms completely detached from the land (barrier islands)*

---

* A recent survey has shown that simple spits and baymouth barriers are the most common forms almost everywhere in Canada.
1. A winged headland is a special case in which a headland has spits extending out from each side.
2. A flying spit is a former tombolo connected to an island that has now disappeared.
3. Looped spits.
4. Double-fringing spits.

Beaches attached to the land at one end. • Sediment moving alongshore often builds spits at the mouths of estuaries, and at other places where there is an abrupt change in the direction of the coast. Most spits tend to extend the original line of the coast, thereby eliminating or reducing irregularities (Fig. 9.10 a,b,c,d). Some, however, project out from a coast towards the ocean, then turn to trend approximately parallel to it (Fig. 9.10 e). A spit is described as recurved if its end curves strongly away from the incoming waves (Fig. 9.10 f), and compound recurved if there are a number of successively landward deflected termini along its inner side (Fig. 9.10 h, g). The term 'serpentine' has been used to describe spits with meandering axes produced by shifting currents, which extend them first in one direction and then in another (Fig. 9.10 k, l). The longshore movement of material down each side of an island can produce comet-tail spits extending from its rear. A single spit normally trails behind narrow islands, but spits often extend from each side of wider islands (Fig. 9.10 o, p, q, r, s). Arrows are spit-like features growing seawards from a coast as they are fed by longshore movement of material from either side (Fig. 9.10 i, j).

Forelands are also attached to the land at one end, but unlike spits and arrows, their width is usually greater than their length. Forelands grow out from coasts, tending to make them more irregular. They can develop where two dominant swells oppose each other, as in the sheltered area behind an island (Fig. 9.11 b), or through movement of material from only one side.

Figure 9.10.
Canadian examples of spits and other depositional forms attached to the land at one end.

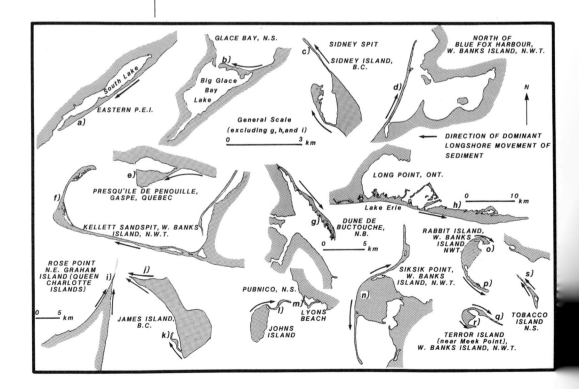

Beaches attached to the land at two ends. • Looped and cuspate barriers are formed when lengthening spits come into contact with the opposite shore, another spit, or an island. Looped barriers develop in the lee of islands when two comet-tail spits become attached to each other (Fig. 9.11 e, s). Cuspate barriers (also known as cuspate forelands) are similar to forelands, except that they enclose lagoons or swampy areas (Fig. 9.11 c, d). They may eventually become forelands, however, if their interiors are drained and filled. Cuspate barriers can develop where longshore movement of sediment is mainly from one direction, as when a looped spit curves back onto the land, or, as in the case of double-fringing spits, where there are opposing directions of longshore movement. Double-fringing spits are formed by the attachment of two spits, or by tombolos growing out to islands that later disappeared (Fig. 9.10 m and Fig. 9.11 f, g, h, i, w).

Single (Fig. 9.11 a, u) or double (Fig. 9.11 r, t) tombolos connect islands to the mainland or islands to islands. Tombolos develop on the lee side of islands because of the shelter provided from strong wave action, and the refraction and convergence of the waves behind islands. Narrow tombolos form where comet-tail spits on the lee side of islands come into contact with the mainland. Y-shaped tombolos develop where comet-tail spits coalesce with cuspate forms growing out from the mainland (Fig. 9.11 q), or where a cuspate barrier is extended landward or seaward (Fig. 9.11 p, v).

Barrier beaches close or almost close off bays and inlets. They can form

Figure 9.11.
Canadian examples of barriers, tombolos, and other depositional forms attached to the land at two ends.

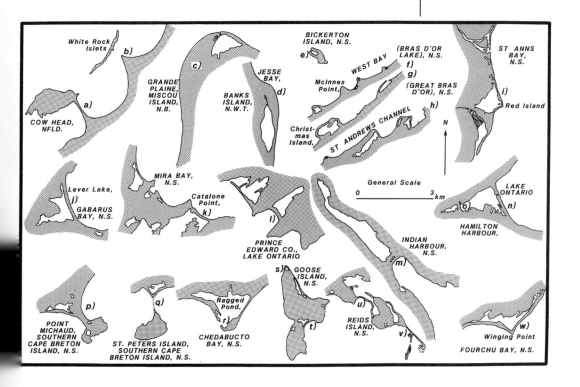

at the mouth (Fig. 9.11 j, k), close to the back or head (Fig. 9.11 n), or at some point within the central portion of bays (Fig. 9.11 m, l). Barrier beaches are produced by single spits extending across bays, or, because of complex patterns of wave refraction, pairs of converging spits built by opposing longshore currents. It has also been suggested that barrier beaches can be built by sediment driven into bays by wave action, independent of longshore movement.

Beaches completely detached from the land. • Barrier islands are long, narrow beach forms completely separated or detached from the mainland (Fig. 9.12). They enclose estuaries, embayments, or narrow lagoons, which are connected to the open sea through channels or tidal inlets between the islands. Some portions of long barrier island chains may actually be large spits or barrier beaches that are still attached to the land at one or both ends (e.g. Fig. 9.10 g). Some barrier islands are the detached portions of long spits, but others have been attributed to the effect of rising sea level at the end of the last glacial stage. It has been suggested that barrier islands developed on the foundations provided by former **dunes**, storm ridges, and berms at the back of beaches, while the lagoons were created when the rising postglacial sea flooded the lower land behind. Alternatively, barrier islands may have been formed out of the sediment driven landwards by wave action as sea level rose.

Figure 9.12.
Barrier island chains in the southern Gulf of St Lawrence.

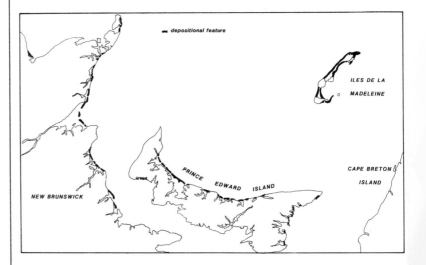

Canadian examples. • Depositional features in eastern Canada are best developed and most numerous where the coast experiences fairly rapid erosion. They are therefore prominent features of the shores of the weak Carboniferous and Triassic rock lowlands, and especially coasts consisting of **unconsolidated** glacial deposits (Johnson 1925, Taylor et al. 1985).

The coast bordering the Carboniferous, and to a lesser extent the Triassic, plains and lowlands of eastern Canada are rich in most types of depositional feature (Fig.1.10). Barrier islands, with associated barrier beaches and spits, have developed across the structurally controlled estuaries and embayments of the southern Gulf of St Lawrence, off eastern New Brunswick, and along the northern shore of Prince Edward Island (Fig. 9.12). Most of these islands are migrating landwards (McCann and Bryant 1972). Spits, forelands, and tombolos are common on the sheltered shore of New Brunswick behind Prince Edward Island, and on the Magdalen Islands, which are connected to each other by a series of tombolos, spits, and barrier beaches (Fig. 9.12) (Owens and McCann 1980). Erosion and longshore movement of sediment on either side of King Head at Merigomish, in northern Nova Scotia, have produced a winged headland, with a spit extending from one side and a barrier beach from the other. The shores of Chaleur Bay also have a collection of simple, hooked, compound recurved, and serpentine spits, and on Miscou Island, at the entrance to the Bay, a series of beach ridges records the stages in the growth of the Grand Plaine cuspate barrier (Fig. 9.11 c).

The Bras d'Or Lakes on Cape Breton Island have one of the most remarkable assemblages of depositional forms in Canada (Fig. 9.11 f, g, and h). Wave action in the long, narrow channels is naturally limited to two opposing directions, providing ideal conditions for the development of forelands and cuspate barriers. Tombolos, spits, cuspate barriers, and forelands line the southeastern shore of the St Andrews Channel, for example, and there are looped spits on the lee of the islands in this area, including one on MacDonald Island at the western end of the St Patricks Channel. Other islands have long comet-tail spits extending from their lee sides.

Shores consisting of easily eroded, unconsolidated glacial deposits, including drumlins, moraines, and sandy **glaciofluvial** outwash, are not widespread in eastern Canada. Where they do occur, however, rapid erosion provides large amounts of sediment. At five sites in Atlantic Nova Scotia, the retreat of the **till** bluffs averages about 0.9 m per year, with a maximum on one exposed headland of 3.3 m. Calculated rates of erosion on the till bluffs in Northumberland Strait and in the Bay of Fundy range from 0.3 to 0.5 m per year (Taylor et al. 1985). An outwash deposit provided the sediment for spits that have almost closed the mouth of Advocate Harbour in the Bay of Fundy (Swift and Borns 1967). These spits, built by longshore drift from opposing directions, have their counterparts in similar raised features at the back of the modern shore, although they were probably deposited when the tides in this area were about one-third of their present level (Fig. 9.13).

Numerous poorly sorted pebble and sand spits, barrier beaches, and single and double tombolos connect the partially drowned, wave-eroded drumlins of Nova Scotia. There are some examples in Mahone Bay near Lunenburg, although conditions are more favourable in Halifax Harbour and on the more exposed coast to the east. Rapid morphological changes

Figure 9.13.
Advocate Harbour, Nova Scotia (Swift and Borns 1967).

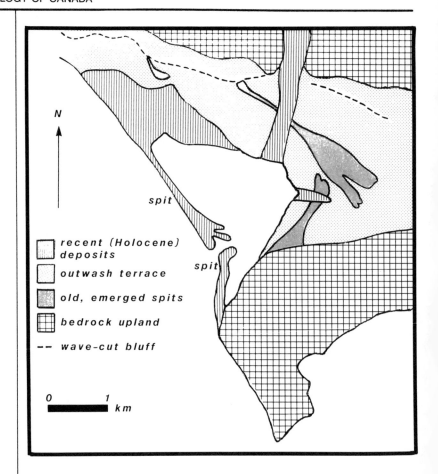

reflect rising sea levels and variations in the supply of sediment (Carter et al. 1989). Some tombolos, which were once attached to rapidly eroding drumlin islands, are now flying spits extending out into open water. Barrier Beach in Halifax Harbour was once an excellent example. It probably developed as a tombolo attached to islands of glacial **debris**, but the islands had been reduced to rocky shoals by 1925 (Johnson 1925), and continued erosion has now destroyed the spit. Cow Bay Beach, a bayhead barrier beach northeast of the entrance to Halifax Harbour, may also have developed as a tombolo connecting a series of drumlins. The drumlins have subsequently been removed by wave action, however, leaving behind deposits of cobbles and boulders in two areas. Removal of large amounts of beach sediment between 1954 and 1971 resulted in considerable erosion of this beach (Fig. 9.14). Although material is no longer being removed, the beach was still retreating in 1981, which suggests a lack of available sediment in the coastal system. There is also a well-developed drumlin coast in southeastern Cape Breton Island (Wang and Piper 1982). This area has good examples of drumlins connected to the mainland or to other drumlins by double tombolos, and Winging Point is a beautifully symmetrical example of a pebble double tombolo or double-fringing spit; it is connected to a rocky reef that is the remnant of a drumlin built around a rock core (Fig. 9.11 w).

Some of the best depositional features in British Columbia are also derived from glacial sediments. They occur on the Nanaimo Lowlands along the eastern coast of Vancouver Island, and on the nearby islands. The features in these areas include a variety of simple and complex recurved spits, arrows, tombolos, and barrier beaches (Fig. 9.10 c, j). A large arrow has also been built from glaciofluvial sediments on the northeastern tip of the Queen Charlotte Islands (Fig. 9.10 i).

There are some excellent examples of depositional features on the shores of the lower Great Lakes. They include the three large, partially moraine-based spits on the northern shore of Lake Erie – Point Pelee, Rondeau, and Long Point (Fig. 9.10 h) – the compound recurved Toronto Island spits, and the barrier beaches at Wasaga Beach on Georgian Bay, Lake Huron, and at Burlington and Dundas (Fig. 9.11 n, o) and on Prince Edward Peninsula (Fig. 9.10 l) at either end of Lake Ontario.

Depositional features are far less important elements of the coasts that border the resistant rock uplands of eastern and western Canada. Wave erosion is slow in these areas, and those features that do occur are probably largely composed of glacial debris, or sediment eroded from weaker rocks in adjacent regions. Pocket beaches of sand, pebbles, or boulders at the head of coves are the most common type, although there are spits, tombolos, gravel barrier beaches, and cuspate barriers in some areas, particularly on the eastern coast of the Gaspé Peninsula (Fig. 9.10 e).

Complex spits are absent on the rocky coasts of the central and eastern Arctic, and even where there is abundant sediment, ice prevents it from being moved effectively along the coast. High **shingle** barrier beach ridges were built across the mouths of many inlets, however, while the land was rising, and there are some simple shingle spits and tombolos on most of the limestone coasts. There is a significant ice-free period of open water in the summer in the western Arctic, around the Beaufort Sea. This area is transitional between the storm-wave environments of southern Canada and the ice-dominated areas of the High Arctic. It experiences far more wave action than elsewhere in the Arctic, particularly during storms, and the ice-rich sand and gravel bluffs of the Arctic Coastal Plain provide abundant sediment. Depositional features can therefore develop very quickly in this area. At Sachs Harbour in southwestern Banks Island, for example, two spits were extended by 400 and 600 m between 1950 and 1979. Material produced by rapid erosion of the bluffs, by slumping and thermal melting of ground ice, has built sand and gravel spits, barrier beaches, and barrier islands throughout this area. Among the many depositional features on the western coast of Banks Island, one might note the beautifully recurved Kellett sandspit and the winged headland of Siksik Point, with its long spits extending out from either side (Figs. 9.10 f, n, and 9.15).

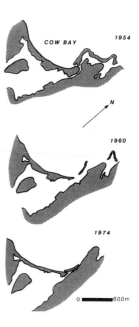

Figure 9.14.
Erosion of Cow Bay, Nova Scotia (Taylor et al. 1985).

COW BAY   1954

N

1960

1974

0 ▬▬▬ 600m

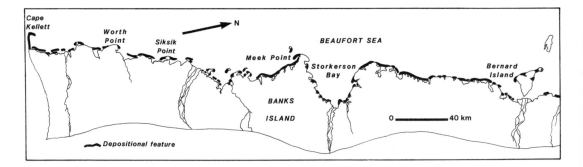

Figure 9.15.
Spits, barriers, and related depositional features on the southwestern coast of Banks Island.

Coastal Sand Dunes

Dunes, or hills of fine, wind-blown sand, are found at the back of beaches in most of the lowland regions of southern Canada. They are well developed in Ontario around the Great Lakes and on the coasts of Hudson and James Bays, as well as along the uplifted shores of ancient lakes and seas (Martini 1981b). There are also dunes on the barrier islands of the Maritime provinces, which attain heights of up to 15 m in northern Prince Edward Island. Although there are small groups of dunes in many parts of coastal Nova Scotia, they are extensive only in the southwest, around Port Mouton and Cape Sable. Dunes are also extensive on Sable Island, a small island of reworked glacial and glaciofluvial sand about 200 km southeast of Nova Scotia. Most of the dunes, which are in the form of two ridges, are between 5 and 15 m in height, but some reach a maximum of about 26 m.

Although there are many different kinds of coastal sand dune, two types are particularly common. Transverse dunes are sandy ridges running approximately parallel to the coast. They sometimes form on top of raised beaches, although their occurrence is usually the result of specific patterns of wind movement. Parabolic or u-dunes have a crescentic plan shape, with their horns facing seawards, towards the onshore winds. They can develop through the **deflation** of transverse dunes, and they are therefore often superimposed upon them.

Vegetation plays a crucial role in encouraging deposition and the continued growth of dunes, and in stabilizing and preserving their form. The types of plants that colonize a dune field depend upon climate and numerous other local factors, but one can usually recognize a marked transition in plant communities as one moves inland. Several major communities have been recognized at Brackley on Prince Edward Island (McCann and Bryant 1972):

(a) The bare, mobile sand of the initial or embryonic dunes is progressively occupied by the marram grass community. Although marram itself plays an important role in the growth and eventual stabilization of these dunes, the community also includes many other species.

(b) A thin, discontinuous layer of litter and humus forms on dry dunes, and as they begin to be fixed, the marram grass community is replaced by a shrub community containing a variety of species. The growth of lichens and mosses further helps to fix the sand and prepare it for colonization by white spruce-bayberry forest and other less specialized species.

(c) Where depressions extend down to the water table, the occupational sequence begins with rushes, and is eventually followed by shrubs, willows, grey birch, and herbaceous species.

Disruption of the protective vegetational cover on a dune exposes the sand to the wind, resulting in the formation of spoon-shaped depressions or blowouts. Vegetation can be damaged or destroyed by waves washing over the dunes, fire, animal digging and grazing (including the excavation of sunning pits by polar bears in the Hudson-James Bay area) and logging, trampling, and other human activity. Recent damage to dunes and the formation of blowouts has resulted from the use of all-terrain vehicles; high recreational use has damaged dune vegetation in parts of southern Canada, causing severe **degradation** of dune fields. Attempts are being made in some areas to reclaim and rebuild the dunes by fencing-off dune areas, building walk-ways through the dunes to the beach, planting vegetation to trap and stabilize wind-blown sand, and using fences as sand traps.

Mudflats and Salt Marshes

Mudflats and salt marshes develop in tidal bays, estuaries, and other areas that are protected from strong wave action, often in the shelter provided by barrier beaches, barrier islands, and spits.

It has been estimated that there are about 280 km^2 of salt marshes and associated bogs in the upper portions of the Bay of Fundy. In the Minas Basin, the middle and lower intertidal zone consists of sand flats up to 5 km in width. The fast tidal currents are responsible for most sand transport. Sand bars, up to several kilometres in length, develop parallel to the currents, and they are covered by smaller dunes and sand waves (Knight 1980). Mudflats replace the sand flats towards the head of the basin, and in bays, estuaries, and other more sheltered places where the tidal currents are weaker. Deposition of the fine-grained, suspended sediments takes place during the slack period at high tide. There are mudflats in the upper parts of the intertidal zone, but the uppermost intertidal and supratidal areas are occupied by salt marshes crossed by deep, muddy creeks. There are also extensive marshes in the sheltered areas of Chignecto Bay, particularly at the head of the Cumberland Basin. The Acadian French began to dyke this marsh in 1670, however, and only a small area now remains in its natural state.

As noted previously, the salt marshes of the St Lawrence Estuary are severely eroded by ice during the spring breakup. Pans or water-filled depressions are formed where sections of the organic cover are raised by ice blocks or floes during high spring tides, and then carried away by the **ebb tides**. Ice erosion is also a prominent feature of the extensive mudflats and salt marshes of Hudson and James Bays. Mud flats in this area are

Figure 9.16.
The Fraser Delta (Clague et al. 1983). (ka BP: thousands of years before present.)

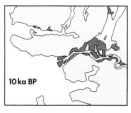

10 ka BP

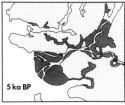

5 ka BP

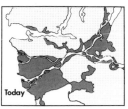

Today

☐ Pre-existing land
▓ Floodplains, fans, and peat bogs

severely eroded by ice, forming grooves up to 2 km in length, and circular or subcircular basins a few metres in diameter; this erosion is usually accomplished when blocks of ice are carried seawards by the ebb tide. There are also extensive salt marshes in Labrador, although they are usually less than 200 m in width, compared with several kilometres in the James Bay Lowlands.

Deltas

Deltas are **alluvial** deposits formed where rivers flow into standing bodies of water. They usually develop where rivers supply fairly large amounts of sediment, and wave energy is not too high. Fiords provide suitably sheltered environments for the formation of deltas on the Pacific, Atlantic, and Arctic coasts of Canada. About 188 major deltas have been identified in the fiords of British Columbia, for example, compared with only about 41 on the open ocean coast (Clague and Bornhold 1980). Three types of delta, each with a distinctive morphology, can be distinguished in the BC fiords (Kostaschuk 1987). Most have developed where large streams enter the heads of fiords, but others have formed where large streams enter along their sides, or where streams run down the steep walls of the fiords. The deltas are covered by trees above the high tidal level, but there are salt marshes and sand flats at lower levels. Potentially dangerous and destructive slides and flows can occur on the subtidal portions of these fiord deltas.

Although there are some large deltas at the heads of fiords on the eastern coast of Baffin Island, major deltaic shorelines are generally lacking in eastern Canada. Large deltas have been constructed on the Pacific coast, however, by the Skeena and Fraser Rivers. The Fraser Delta in southwestern British Columbia is much younger than most of the other major deltas in the world, having begun to form only about 8,000 years ago, after deglaciation and local postglacial uplift had been largely completed (Fig. 9.16) (Clague et al. 1983). It developed in a fairly high-energy marine basin with a tidal range of between about 4 and 5 m, and unlike most other deltas, which largely consist of fine-grained silt and clay, half its sediment is sand. About 20 million tons of sediment is transported past New Westminster every year, and then carried across the delta in the **distributaries** of the Fraser River. Most sediment is presently deposited near the mouths of the distributaries on the western side of the delta, where it is gradually being extended seawards. The position of the distributaries has changed through time, as old channels have been abandoned and replaced by new ones. But much of the approximately 1,000 km² of the delta has now been dyked and dredged, preventing changes in the river course and the deposition of sediment by overbank flooding. A high proportion of the delta is therefore receiving no new sediment today and is essentially inactive.

The Mackenzie Delta is about 210 km long in a north-south direction, 64 km wide, and approximately 12,000 km² in area. Although the Peel and Rat Rivers contribute sediment to the southwestern part, the delta is largely the product of the Mackenzie River – the longest river in Canada and one of the ten longest in the world. The delta contains an intricate network of channels and thousands of shallow, interconnecting lakes. Most of these lakes contain water trapped by the higher ground around the channels, although some may be thermokarst features (Chapter 7) produced by the melting of ground ice. Levees (Chapter 8) are poorly developed on the older, higher floodplains of the southern half of the delta, but they are common in the north. Flooding can occur on the delta during breakup, when ice jams develop, and in coastal areas during storm surges. All regions are susceptible to flooding, although it is rare in some of the older parts of the delta.

Rock Coasts

A high proportion of Canada's coast consists of rock that is very resistant to the erosive processes arrayed against it. There is therefore a widespread scarcity of beach sediment, and in many areas the only depositional features are small, stony pocket beaches between resistant headlands, with deltas and mudflats in sheltered areas. Coastal slopes also reflect the strength of the rock. There are steep marine cliffs in the weak **sedimentary** rocks along parts of the Atlantic and Pacific coasts, and in the limestones of the eastern and central Arctic. Steep bluffs are also produced in rapidly retreating unconsolidated deposits in eastern and western Canada, the western Arctic, and the southern Great Lakes. Along much of the Canadian coast, however, glacially sculptured crystalline and other resistant rock plunges or slopes beneath the sea with little sign of modification by wave action. In some areas, such as Cape Breton Island, where wave attack has been a little more successful, the cliffs have composite or hog's back profiles consisting of a short, steep marine cliff beneath a long, sloping **subaerial** surface (Fig. 9.17).

Many mechanisms contribute to the erosion of rock coasts (Trenhaile 1987). Mechanical wave erosion is accomplished by a number of processes, including the compression of air trapped in rock crevices by the incoming waves or swash, and the abrasive action of sand, shingle, and pebbles as they are swept back and forth across rock surfaces. Other mechanisms are important, and sometimes dominant, in particular regions, however, depending upon such factors as the local climate, rock type and structure, and the degree of exposure to vigorous wave action. They include the **corrosion** of limestone and other chemical weathering processes, biological erosion associated with marine micro**flora** and **fauna**, and rock falls, slides, and other types of mass movement. Frost and related processes (Chapter 2) play a particularly important role on Canada's coasts where the rocks are suitable.

Mechanical wave erosion is very sensitive to variations in rock strength as a result of changes in rock type or structure. In much of coastal Canada, marine erosion has been limited by the strength of the rock and, because

Figure 9.17.
The glacial-interglacial origin of composite cliff profiles (Trenhaile 1987).

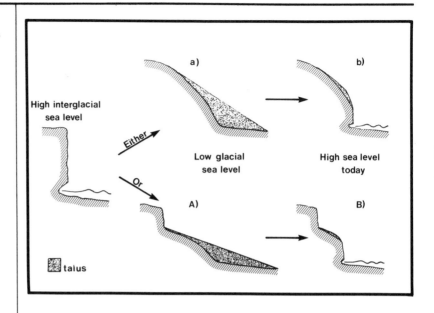

of changes in relative sea level, the short time that the sea has operated at its present level. All that has been accomplished in many areas is the etching out of **joints**, **faults**, **bedding planes**, and other lines of weakness to form small inlets and caves. In less resistant rocks, however, the more effective exploitation of weaknesses has produced caves, coves, blowholes, narrow inlets (geos), stacks, and arches (Fig. 9.18). These features tend to occur together, wherever the rock is well jointed and of sufficient strength to stand in nearly vertical cliffs and as the roofs of caves, tunnels, and arches. There are isolated examples along the Triassic and Carboniferous coasts of the Maritime provinces. An interesting stack has developed in the columnarly jointed **basalts** of Brier's Island near Digby, Nova Scotia, on a coast reminiscent of the better known Giant's Causeway in Northern Ireland. Percé Rock is a long, narrow island cut from a ridge of vertically bedded limestones off the coast of eastern Gaspé, Québec. The island is flanked by steep, possibly fault-controlled cliffs up to about 90 m in height, which were formed by the removal of the weaker rocks on either side. Percé Rock is pierced by one large arch, while a stack at its seaward end probably represents the remnant of another whose roof has collapsed. The mushroom-shaped stacks, caves, and arches at Hopewell Rocks near Moncton in the Bay of Fundy probably represent Canada's best example of an association of erosional features cut along lines of structural weakness.

Horizontal or gently sloping rock surfaces develop at the foot of retreating coastal cliffs (Fig. 9.18). These shore platforms are an expression of the ability of marine and subaerial mechanisms to erode a rock coast, and their best development is therefore in weak to only moderately resistant rocks in fairly vigorous wave environments. There are narrow ledges or platforms in places in the Arctic and subarctic, but they attain widths of several hundred metres or more on southern Vancouver Island and in several parts of eastern Canada (Trenhaile 1978). The type of shore platform that develops appears to be determined primarily by the tidal range. Platforms in the micro- and mesotidal environments of eastern Newfoundland and Gaspé, Québec, are essentially horizontal midtidal surfaces, which terminate abruptly seawards in a low tide cliff or ramp. In the macrotidal Minas Basin at the head of the Bay of Fundy, however, the platforms slope seawards with gradients of 3 to 5°, extending from the high tidal level to below the low tidal level without a marked break of slope.

Figure 9.18.
Erosional features of a rock coast.

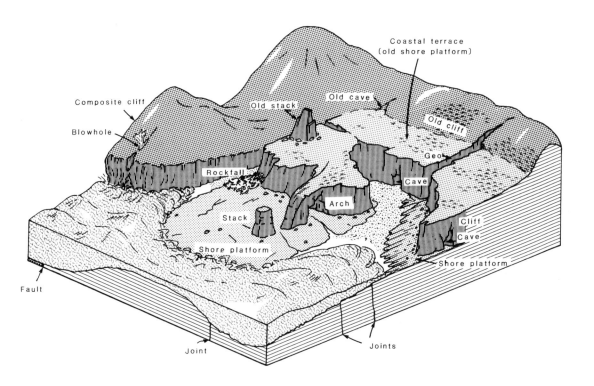

10 | Karst

The term 'karst' is the German form of *kras*, a Slovenian word originally meaning bare, stony ground; Kras is also the name of a rugged region in western Slovenia, near the Adriatic coast of Yugoslavia. Karst refers to terrain where soluble rocks have been greatly modified, above and below the surface, by the dissolving action of water.

Although most rocks are at least slightly soluble in water, extensive karst landscapes usually develop in the carbonate rocks (limestones and dolomites), or in some cases in the evaporites, which include halite (rock salt, NaCl) and the sulphates, anhydrite ($CaSO_4$) and gypsum ($CaSO_4.2H_2O$). Limestones are a complex and varied group of rocks. At least 90 per cent of the mineral content of pure limestones is calcite ($CaCO_3$), whereas at least 90 per cent of pure dolomites is dolomite ($CaMg(CO_3)_2$). A number of terms, including 'magnesium limestone', 'dolomitic limestone', and 'calcitic or calcareous dolomite', are used to refer to carbonate rocks that are intermediate between pure limestone and pure dolomite. Many other geological factors, however, besides the chemical or mineral composition of the rock, play an important role in determining the nature of karst development. These factors include the physical strength and permeability of the rock and the presence and nature of any overlying soil, **sediment**, or non-soluble bedrock.

Simple physical solution of evaporites is very effective. Gypsum and halite, for example, dissolve in the presence of water through simple dissociation. For salt:

$$NaCl = Na^+ + Cl^-$$

Because they are so soluble, karst development in salt, and to a somewhat lesser extent gypsum, is most common where the rocks are covered, and thus protected, by sediments or less soluble rocks. But limestones are not very soluble in pure water, and solution is therefore usually dependent on carbon dioxide from the atmosphere or the soil dissolving in water to form carbonic acid. Solution of limestones in most karst water is the result of the following reactions:

(a) for calcite

$$CaCO_3 + H_2O + CO_2 = Ca^{2+} + 2HCO^-_3$$

(b) for dolomite

$$CaMg(CO_3)_2 + 2H_2O + 2CO_2 = Ca^{2+} + Mg^{2+} + 4HCO^-_3$$

Sulphuric acid produced by the weathering of sulphide minerals and by other means, and organic acids released by plants and animals are also important in some situations.

Landforms

Few geomorphic processes are unique to karst regions, although several operate in unique ways, or with particular effect, in areas underlain by limestones and other soluble rocks (Ford and Williams 1989). Karst development requires at least moderate amounts of precipitation, and fairly soluble, well-jointed rocks. Water is transported underground in the **permeable** rock, and surface drainage may be widely spaced, intermittent, disrupted, or absent. Karst landforms can be classified according to whether they are found on the surface of the rock, within the rock, or where the water eventually flows out of the rock.

Surface Features

Rainfall and snowmelt produce a variety of surface features in karst regions. Limestone and dolomite pavements are roughly horizontal rock benches or plains, often glacially scoured, with major **joints** opened up by solution down to depths ranging between about 0.5 and 25 m. The surfaces (clints) between the open joints (grikes) carry a variety of features known as karren, or lapiés, although they are also found on other types of surface. Solutional etching produces a bewildering array of karren, including sharp pinnacles, ridges, grooves, circular basins, steps, and jagged fissures, ranging in size from a few millimetres up to several metres. Smoother, more rounded types usually develop under a peat, vegetation, or soil cover. The type of karren that develops in any area depends upon such factors as the amount, nature, and distribution of precipitation; the nature, texture, **dip**, and structure of the limestone; the chemical reaction involved; and the effect of former climatic phases.

Solution depressions, known as sinkholes (American) or dolines (European), are among the most typical features of karst regions. They are closed depressions that tend to have a circular or oval plan-shape, and a

Figure 10.1.
Types of sinkhole. The suffosion doline is the most common type of karstic closed depression in Canada.

a)

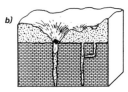

b)

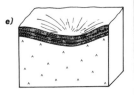

c)

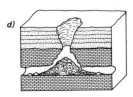

d)

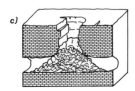

e)

a) *solution sinkhole*
b) *suffosion sinkhole*
c) *collapse sinkhole*
d) *collapse through insoluble rock overlying limestone ('covered karst')*
e) *subsidence sinkhole*

bowl, conical, or cylindrical sectional-shape. **Cockpits** are irregular, star-shaped hollows in the humid tropics. Sinkholes range from a few metres up to several hundred metres in diameter, and from a few metres to more than a hundred metres in depth. They can have vegetated or rocky sides, and occur as isolated individuals or in large groups. The term 'uvala' refers to larger, more complex depressions with uneven floors and two or more low points, created by the coalescence of several sinkholes, or through the collapse of the roof of a subterranean **stream** system. Sinkholes can be formed in several ways, including solution acting from the surface downwards along fractures or joints; piping (washing) or slumping of glacial **drift**, **alluvium**, or other sediments into solutional openings in the karst rock below (suffosion); collapse of bedrock into an underlying solutional cavity; and subsidence of insoluble strata as a result of gentle solution of underlying gypsum, salt, etc. (Fig. 10.1). In cold climates, sinkholes are widened and deepened by solution beneath snowbanks, as well as by frost and other nivational processes.

The largest karst depressions, which are known by the Slavic term 'poljes', are typically elongated basins with flat floors and steep rock walls. Some poljes are completely surrounded by limestones, while others have formed at the contact between limestones and impermeable rocks. Non-calcareous material accumulates in the poljes, and their floors are usually covered in alluvium, or glacial or periglacial deposits. Hums or residual hills of limestone may protrude through the surface of the plains formed in these sediments. Many poljes flood during wet periods, when the sinks are unable to accommodate the surface drainage, and **groundwater** issues from rock fissures. Poljes are generally thought to be produced by solution, probably through lateral undercutting of the surrounding slopes during floods.

Running water tends to disappear underground in karst regions because of the permeability of the limestones. Streams flow into vents, which are known as sinks in America, swallow holes or swallets in Britain, and ponors in Slovenia. Some streams become progressively drier as the water is gradually lost along its course down joints and **bedding planes**. Others disappear more abruptly, flowing into more or less horizontal caves, vertical or steeply sloping shafts, or holes in drift or alluvium overlying limestone.

The loss of water at some point along a stream course reduces the amount of solution and mechanical erosion occurring downstream. The difference in the **erosion** rate above and below a sink can cause a step to develop, with the riser facing upstream. This becomes an increasingly formidable obstacle to downstream flow. A valley is defined as being half-blind if stream flow can continue beyond the step into the lower part of the valley during snowmelt or heavy rainfall, when the sink is unable to accommodate all the water. In blind valleys, which terminate in a step or cliff ranging from a few metres to hundreds of metres in height, all the stream flow is diverted underground, through a series of sinkholes or through a cave at the base of the cliff. Dry valleys do not have stream channels in their floors. Some are the continuation of stream valleys beyond the furthest points now reached by surface drainage, but the origin of branching or dendritic dry valley systems is more problematic. It has been suggested that they became dry in some areas when there was a change of climate, or when the deepening of major stream valleys lowered the **water table** in the tributaries. Others probably developed on non-karst rocks, and were later superimposed on the limestones below.

Streams rising on impermeable rocks sometimes flow across limestone regions, cutting allogenic (through) valleys. Only fairly large streams have enough volume to cross karst regions, however, and even they may be dry for part of the year; smaller streams lose all their flow and become blind valleys. Allogenic valleys are often in the form of narrow, steep-sided canyons or gorges. Gorges are more common in limestones than in other rock types. This is because the infiltration of water into the ground and the reduced surface runoff make it difficult for slope processes to widen the valleys and reduce the steepness of their sides. Effective undercutting at the base of limestone slopes by meandering streams may also contribute to their steepness, while the valleys are deepened by solution and abrasion in the stream channel. Gorges or portions of gorges can sometimes develop through the collapse of the roof of an underground cavern, and if a portion of the roof remains in place, the gorge may be spanned by a natural bridge or arch.

Caves

The term 'cave' is loosely defined as an accessible passageway or space, but there are also innumerable smaller tubes and fissures that perform the same hydrological function. Caves range from simple systems consisting of short passageways, open shafts, and single rooms, to complicated three-dimensional systems with shafts, chambers, rooms linked by passages, and halls of different shape and size. Streams flowing underground at different levels within the rock may cross each other, and they can flow in directions that are contrary to the slope of the surface above. In southern British Columbia, for example, some groundwater systems flow beneath the continental divide to springs that issue in Alberta.

The water table in karst regions tends to be more irregular than in other strata, and it may not be definable in young mountainous regions. Nevertheless, it is usually possible to recognize the occurrence of a vadose zone above the water table and a phreatic zone below. The larger spaces in the rock in the vadose zone contain air, and water flows downwards under gravity, whereas in the phreatic zone, the water flows under pressure in the water-filled **conduits**. In an intermediate zone between the vadose and phreatic zones, flow is sometimes gravitational and sometimes under pressure, according to fluctuations in the water table. Ford and Ewers (1978) have suggested that caves develop (Fig. 10.2) in

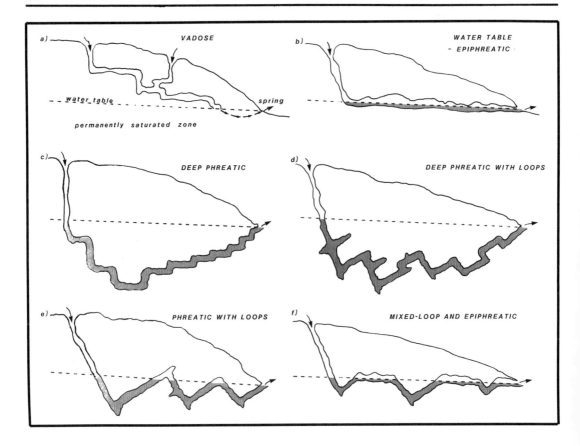

Figure 10.2.
Phreatic, vadose, and water-table caves (Ford and Ewers 1978).

(a) the unsaturated vadose zone, which may be dry in places or at certain times;

(b) the intermediate zone, or just beneath and parallel to it (water table, shallow phreatic, or epiphreatic caves); or

(c) the deep phreatic zone, where the cavities are permanently filled with water. Looped cave systems can penetrate to considerable depths in karst rock.

It has been suggested that two other types of cave are intermediate between types (b) and (c). If the frequency of the rock fissures is greater than in type (c), phreatic caves may develop a sequence of downward loops. Other caves consist of sections of horizontal epiphreatic passages connecting phreatic loops.

Since caves can develop in any of the hydrological zones, cave systems may be a combination of vadose sections in the higher areas and water table or deep phreatic sections in the lower portions. Solutional features are produced in the vadose zone by water flowing under gravity through the rock, along the steepest available course. Less steep courses may also be used, however, if the steepest path cannot accommodate all the flow. Vadose passageways are often canyon-like, with entrenched floors trending continuously downwards. Waterfalls often connect different levels, and deep, well-like shafts develop where the water flows down vertical or steeply dipping fractures. Water can be moved under **hydrostatic pressure** in phreatic systems, and it has no particular tendency to follow the steepest path. Phreatic passageways tend to be circular or elliptical in shape, and they have gentle overall gradients, possibly with upward- or downward-trending segments of steeper slope. As cave systems develop, older passages are abandoned and newer ones created at lower levels to take their place; this results in the modification of older phreatic groundwater systems by vadose flow.

The development of karst cave systems requires rock that is strong enough to stand in the roof of passageways, and pure enough that there are no insoluble residues to block narrow conduits. Bedding planes, joints, and other planes of weakness allow groundwater to penetrate into soluble rock, but cave development is inhibited if they are too close together. Although they are highly soluble, many evaporites are too weak to support the roofs of extensive cave systems. Nevertheless, some of the largest caves are in gypsum. The development of karst caves also requires a supply of water, and sufficient topographic **relief** to generate a hydraulic gradient to move it about. If these conditions are satisfied, the type of cave that develops depends upon the density of the fissures, the structure and lithology of the rock, and the local relief between the sinks and the springs. Segments of caves often follow bedding planes, joints, or **faults,** or the intercepts of these fissures. Groundwater can flow up or down the dip of the rocks, or along the **strike**. For example, the sawtooth profiles of many phreatic systems are the result of groundwater flowing upwards and downwards along the joints and bedding planes.

The precipitation of calcite and other minerals in caves produces a tremendous variety of speleothems, or cave deposits. Stalactites are formed by drips on ceilings, whereas stalagmites develop where the drops fall to the floor. Columns are created when stalactites and stalagmites join together. Other types of speleothems are the result of water trickling down the walls of caves, and wide films of water over their floors. Speleothems can develop in active caves as long as they are not permanently full of water. But they are frequently damaged and covered in mud by running water, and their best development is therefore in abandoned caves, where they can eventually completely block the passages. Other cave deposits include sediments deposited by mass movement, wind, streams, and lakes, as well as by ice in cold climates.

Cave deposits can be dated using **Carbon-14,** and particularly with the $^{230}Th/^{234}U$ ratio in the uranium radioactive decay series. Most calcite speleothems contain uranium derived from the weathering of soils, and although present only in very small amounts, it is sufficient for dating purposes. The current limit of the uranium-thorium technique extends back to 350,000 to 600,000 years, but as other potential methods are perfected, it should become possible to date calcite deposited within the last one and a half million years; Canadian workers continue to be the world leaders in the dating of cave deposits using these and other methods. The importance of dating cave deposits transcends its obvious application to karst geomorphology. For example, the U-series method has been used to date speleothems in relict phreatic caves perched on the sides of valleys in the Crowsnest Pass and Columbia Icefield area of the southern Rockies (Ford et al. 1981). This has provided estimates of a maximum rate of valley downcutting of between 0.13 and 2.07 m per thousand years. These values suggest that the age of the average relief of 1,340 m between the valley floors and the crestlines in this area is between 1.2 and 12 million years. Furthermore, as speleothems cannot develop during cold periods, when the **percolating** water freezes, dating can determine when these periods occurred. This technique has been used in Canada and elsewhere to determine the occurrence of **glacial** and **interglacial** stages. Speleothems in the Nahanni, Crowsnest, and Castleguard areas of the Mackenzie and Rocky Mountains, for instance, suggest that interglacials occurred from 15,000 years ago to the present, between 90,000 and 150,000, 185,000 and 235,000, and 275,000 and 320,000 years ago, and 350,000 or more years ago. On Vancouver Island, speleothems were deposited in the **Holocene**, and in the Olympia period of deglaciation between 67,000 and 28,000 years ago. A further source of important information is provided by the oxygen isotopic ratio $^{18}O/^{16}O$ in speleothems (Chapter 4).

Springs

Groundwater flows out of karst rocks in springs. The distinction has been made between springs that are largely fed by conduit flow where surface streams have disappeared underground (resurgences), and those that are mainly fed by the more diffuse flow of water seeping down through the rocks (exsurgences). Conduit-fed springs react to rainfall faster, and with much greater maximum flows, than springs fed by percolating water, although most springs display mixed behaviour in well-developed karst. While some springs are simply fed by gravitational flow from downward-flowing groundwater, others are supplied by groundwater welling or flowing upwards under hydrostatic pressure (Vauclusian springs), sometimes through alluvium or drift, or into the bottom of pools. Some springs function only during flood conditions, when the normal spring is unable to accommodate all the groundwater supplied to it. Another type of intermittent spring, known as an estavelle, is often found in poljes, where they can function as either sinks or springs according to the season. Cliffs may form arcuate or cirque-like alcoves around the head of springs, as a result of the progressive collapse of caves and headward sapping. Retreat of large springs as they cut back into the limestone mass produces steep-sided and flat-bottomed pocket valleys.

The Effect of Glaciation

Glaciation has had a profound effect on the development and nature of karst landscapes in Canada (Ford 1983a, 1987). Although glaciation can stimulate karst development in several ways, the overall effect in Canada has more often been to destroy karst landforms or to inhibit their development. As a result, karst systems are not normally as well developed as those in the unglaciated regions of the United States.

Meltwater streams flowing from an ice terminus usually carry large amounts of sediment, which interfere with karst drainage. Low levels of carbon dioxide in glacial ice further reduce the solutional potential of the meltwater beneath the ice. Meltwater is therefore probably incapable of producing karst in unmodified limestones, but it may be able to erode vertical shafts and other short, parasitic passages in pre-existing karst networks.

Glacial dissection of alpine regions by cirque and valley glaciers breaks up subterranean drainage systems. In nonglaciated areas, springs usually issue from carbonate rock on the floors of valleys, but in alpine Canada and similarly glaciated regions, they often hang on the deepened valley sides, or rise through glacial deposits from deeply buried rock outlets (Fig. 10.3). Glacial erosion can also remove karren from rock surfaces, although larger and deeper features such as sinkholes generally persist in a modified form. The efficiency of glacial erosion, however, is partly dependent upon the temperature of the ice at its base (Chapter 4). For example, while limestone pavements were removed on Anticosti Island and on the Bruce Peninsula, a preglacial dolomitic pavement in the Winnipeg area, along with underlying cave systems, may have survived beneath cold-based ice that was frozen to its bed (Ford 1983b). This latter pavement is now beneath a thick cover of drift, which has enhanced the quality of the karst **aquifer**.

Sinkholes, poljes, and valleys can be filled and obscured by glacial and **glaciofluvial** material. Medicine Lake in the Rockies is thought to be a karst polje that was infilled with glacial sediment, and has since been only partly re-excavated. Karst caves can be injected and blocked by glacial material or by the ice itself. Large cave galleries beneath the Columbia Icefields have been completely filled with glacial till. In the Goose Arm area of western Newfoundland, the presence of large, poorly drained sink-

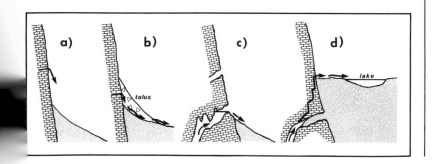

Figure 10.3.
Types of spring in the karst of the Rocky Mountains (Ford 1979).

holes and disordered groundwater drainage suggests that the karst aquifer has been clogged by clay injected into conduits and sinks by subglacial water (Karolyi and Ford 1983).

Renewed karst development in the Goose Arm area, and its development elsewhere in Canada, is inhibited by the presence of carbonates in the overlying glacial deposits. This material can exhaust the solvent capacity of the percolating groundwater before it reaches the underlying bedrock. This is the main reason why there has been only limited karst development in many lowland areas in Canada. The presence of karst in such places as Anticosti Island in the Gulf of St Lawrence, and the Bruce Peninsula and a few places in the Hudson Bay Lowlands of Ontario, can be attributed to a locally thin or absent cover of drift. Elsewhere in Ontario, however, only one or two metres of glacial sediment have afforded complete protection to the underlying limestones and dolomites for the last 11,000 to 14,000 years.

Figure 10.4.
Distribution of karst rocks in Canada (Ford 1983a).

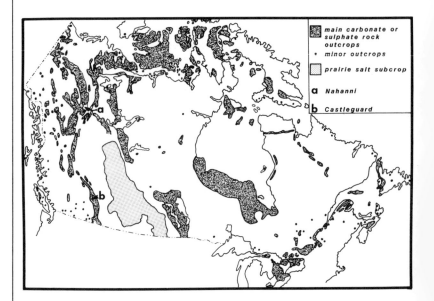

Karst in Canada

There are widespread outcrops of carbonate and sulphate rocks in Canada, in temperate, Arctic, and alpine environments, although they are often buried beneath glacial deposits (Fig. 10.4). Limestone and marble outcrops cover an area of about 5.7×10^5 km², dolomite about 6×10^5 km², and sulphate rocks about 8×10^4 km². There are also large deposits of salt beneath the sedimentary rocks of much of the central prairies.

The best alpine karst in western Canada is found in well-bedded and massive limestones (Ford 1979). Nevertheless, karren, small sinkholes, and short underground stream courses have developed in thinner limestones and dolomites contained within formations that largely consist of less soluble rocks. In Glacier National Park in British Columbia, for example, a stream from the Tupper Glacier plunges into a spectacular sink, and then flows underground for about 2 km along a narrow band of impure limestone, surrounded by an enormous thickness of shale, slate, and grit (Ford 1967). Caves, possibly a kilometre or more in depth, have also developed in thin limestone formations in the Mount Robson area and in the Bocock Peak area of the Foothills further north.

Carbonates are the most common mountain-building rocks in the Foothills of the Rocky Mountain system, and they are also common in the main ranges. Karst occupies large portions of the most scenic areas, including most of Waterton, Banff, and Jasper National Parks. Many features in these areas, some of which are well known to tourists, have a karst origin, including the Ink Pots group of springs and ponds near Banff, the Banff hot mineral springs, the Maligne and Johnston Canyons, the Athabasca Falls, and springs, including those at Crowsnest and Castleguard, that discharge the underground flow of large streams. An enormous subterranean stream-cave system also feeds the Maligne River from Medicine Lake, in and below Maligne Canyon. About one hundred caves have so far been explored and mapped in the Rockies. They include the 536 m-deep Arctomys Cave in Mount Robson Provincial Park, which is the deepest known cave in North America, and the 20 km-long Castleguard Cave in Banff National Park, which is the longest in Canada. Because of steep rock dips, many of the caves in the Rockies are of the sloping phreatic type, with groundwater passing up and down along their courses.

Glaciation has played an important role in the development of karst in the Rocky Mountains. For instance, in the last 3 to 4 million years the Crowsnest Pass area has experienced many glacial cycles, which have resulted in karst modification of glacial landforms, and glacial modification of karst landforms (Ford 1983c). Older cave systems have been **deranged** by glacial action and invaded by younger streams that have opened new courses in them. Many of the cirques cut in limestones, as well as some glacial lakes, are partially or entirely drained underground through sinkholes in their floors. Indeed, cirques are so common in limestone regions that it has been proposed that some may be preglacial sinkholes, which were modified by glacial erosion.

The Castleguard area in Banff National Park is probably the best example of an alpine, subglacial karst landscape known today (Fig. 10.4) (Ford 1983d). It has developed in thickly bedded limestones with a well-spaced, long, and deep joint system. A large cave lies beneath this mountainous area, which contains the highland ice caps of the Columbia Icefield, as well as numerous valley and cirque glaciers. Karst and glacial erosion have therefore proceeded in tandem in this area. There are karren on the surface of most of the exposed carbonate rocks, and some are developing in areas that were uncovered from beneath the ice only 30 to 60 years ago. Other small-scale features are associated with pressure-induced melting at the base of the ice, which has allowed some local solution to take place. Upon refreezing, the calcite is precipitated in the areas of lower pressure on the lee side of small obstructions on the ice bed. The main surface features, however, are sinkholes, ranging from narrow, vertical shafts elongated along joints, to circular, elliptical, and funnel-shaped forms. Some of these sinkholes appear to have developed beneath ice at the **Wisconsin** glacial maximum, or even earlier.

Unlike most of the caves known in the southern Rockies, the Castleguard Cave seems to have developed close to the water table, although the main passageways were already drained and relict by at least 750,000 years ago (Fig. 10.5). These passageways were abandoned when younger systems, which are still active today (though inaccessible), developed at lower levels. Meltwater drains into sinks beneath the ice, and then into the caves, eventually reappearing as springs. Over one hundred perennial, seasonal, and episodic karst springs have been identified in this area.

Although large portions of Arctic Canada are underlain by water-soluble rocks, well-developed karst features are generally thought to be quite rare (Bird 1963). Frozen ground in the continuous permafrost zone (Chapter 7) restricts underground flow and solution on the Arctic Islands. It is therefore assumed that karst development is restricted to the seasonally thawed active layer, where there may be some limited solution along the joints and fissures of limestone pavements, particularly under a cover of snow. Nevertheless, the discovery of sinkholes, sinks, springs, and caves in the northern Yukon,

Figure 10.5.
*The Castleguard Cave
(Ford 1983d).*

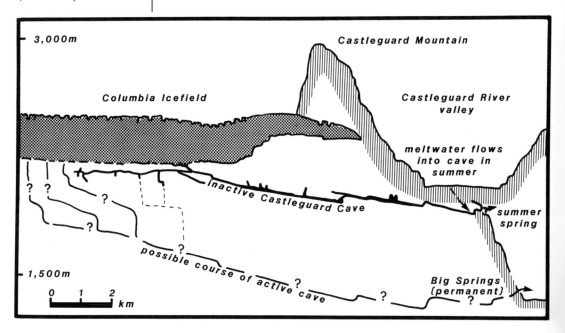

on the edge of the continuous permafrost zone (Cinq-Mars and Lauriol 1985), shows that more work is needed to determine the extent and conditions necessary for deep groundwater circulation in permafrost regions.

There are extensive underground drainage systems in areas of discontinuous permafrost in the subarctic. Indeed, possibly the best karst landscape in Canada is in the Devonian limestones, and to a lesser extent in the underlying dolomites, of the Nahanni region of the southern Mackenzie Mountains (Fig. 10.4) (Brook and Ford 1974). A dense assemblage of karst forms has developed within an area of about 50 km in length and 6 to 12 km in width, including caves, sinkholes, poljes, karst towers, natural bridges, and limestone pavements. The region also contains some of the best examples of limestone gorges in Canada. Four canyons have been incised along the middle section of the South Nahanni River, to a maximum depth of more than 1,000 m, although depths of between 300 and 600 m are most common.

One would normally expect to find only a few small, poorly developed solutional features in cold, dry regions. The remarkable collection of karst landforms in the Nahanni region is therefore somewhat of an anomaly, especially as annual precipitation is only about 520 mm. The area escaped the destructive effects of glaciation in the Wisconsin period, however, and it may have been ice-free for at least the last 300,000 years. Furthermore, rapid upward arching of this mountainous area has bent and fractured the rocks, promoting rapid underground movement of water.

Bare limestone pavements over large parts of the Nahanni region are dissected by deep solution sinkholes, and by joints and other vertical fissures that have been exploited by running water. The enlargement and coalescence of these features along joints and faults have produced spectacular labyrinths, consisting of steep-sided limestone corridors or streets. Intersecting corridors range from a metre or so to more than 200 m in depth, and they can be more than a hundred metres wide and several kilometres long. Although the corridors are primarily solutional in origin, frost action also operates on their walls. The widening and deepening of these corridors eventually produce large, irregularly shaped open spaces, or karst platea, within the labyrinths, sometimes containing scattered, steep-sided residual towers of rock (Brook and Ford 1978).

Sinkholes up to a kilometre in diameter are widespread in the Nahanni region. They are the result of either solution acting on the limestone from the surface downwards, or the collapse or subsidence of a covering of shale or glacial sediment into an underground cavern. Because the area lies within the discontinuous permafrost zone, ice tends to plug the percolation routes, causing water to be retained on the floors of many sinkholes. Although poljes are usually associated with much warmer regions, there are several well-developed examples in this area. The flooding of these poljes after several days of heavy rainfall demonstrates that the karst is still active and evolving. The active and inactive cave systems of the region have been only partly explored. **Radiometric dating** shows that the greatest development of speleothems near the mouth of one cave took place between 300,000 and 200,000 years ago. While it has generally been too cold for the formation of stalactites since that time, some thin, straw-like features have formed in the last 8,000 years.

There are other extensive karst areas in northern Canada, although they are generally dominated by sinkholes rather than by tower karst and labyrinths, and are often associated with the solution of buried gypsum or salt, rather than limestone or dolomite. There is widespread karst to the west of Great Bear Lake, in the Franklin Mountains, Colville Hills, and the Great Bear Plain. Karst features include disrupted surface drainage, springs, and a couple of dry valleys, as well as some large poljes and numerous funnel-, dish-, or cylindrically-shaped sinkholes. Some of the shallow, saucer-shaped sinkholes may be subsidence features produced by solutional lowering of the underlying bedrock; or they could be collapse features filled with sediment. Most of the sinkholes in this area, however, including some particularly large, steep-sided, and deep examples, were probably formed by collapse, induced by the subsurface solution of extremely soluble salt (halite) and gypsum contained in the rock formations below. A similar theory, involving the underground solution of evaporites, has been used to explain the occurrence of small, round, water-filled sinkholes in northeastern Alberta and the adjacent portions of the Northwest Territories. This area consists of Devonian limestones and dolomites, with substantial amounts of gypsum, anhydrite, and salt buried beneath glacial deposits. The sinkholes are sometimes distributed along a straight or gently curving line, suggesting that they were produced by the collapse of caves. There are solution and collapse sinkholes in Wood Buffalo National Park, but groundwater solution of gypsum beds has also caused some normal faulting and subsidence of blocks of overlying strata.

The solution of evaporites is also important in several areas in southern Canada. Large, buried collapse structures and structural depressions in southern Saskatchewan were formed by the solution of underground salt by ground water. Crater Lake in southeastern Saskatchewan is about 244 m in diameter and 6 m in depth, but many other features produced by collapse do not affect the present ground surface. Other examples of karst development related to buried beds of gypsum have been described in the Kootenay Ranges of British Columbia, where the main features are collapse sinkholes (Wigley et al. 1973), and south of Moncton in New Brunswick, where the features include sinkholes, small dry valleys, and caves (Schroeder and Arseneault 1978). Gypsum outcrops in parts of Newfoundland and Nova Scotia have dense microkarren pinnacles up to 25 m high, and caves, micro-cockpits, and poljes. Solution collapse in central Nova Scotia has been attributed to the presence of beds of anhydrite.

Because dolomite dissolves more slowly than calcite, dolomitic karst is generally more subdued than limestone karst, with shallower and fewer sinkholes and smaller and fewer caves. On the other hand, dolomitic pavements seem to be more common than those in limestone. Karren have developed on dolomitic pavements along the crest of the Niagara Escarpment, from the Bruce Peninsula to the Niagara Peninsula, but their development has been inhibited by a widespread cover of soil or calcareous drift. Karren on the Niagara Escarpment have developed along joints, bedding planes, and lithological features, and geological factors may generally play a more significant role in determining their form and nature than in limestones. Many of the small karst caves in Ontario are on the Niagara Escarpment, and there are some small underwater drainage systems. Beaver Valley near Georgian Bay offers a good example of early karst development, including a set of sinkholes, narrow, poorly developed caves formed along the two major joint sets in this region, and both a perennial and an overflow spring (Cowell and Ford 1975).

Although speleothem dating demonstrates that one cave in Gaspé became dry more than 200,000 years ago, Québec's known caves are generally small and young. They are geologically varied and quite numerous, however, and they occur in several areas, including the Ontario and Québec sides of the Ottawa Valley, along the Gatineau River north of Ottawa, the St Lawrence Valley, Lake St John Lowlands, Gaspé Peninsula, and Anticosti Island, and in the metamorphosed limestones of the Grenville Marbles. Some small caves in the Montreal area were formed in limestones and shales by glacial dislocation or ice push of the upper parts of the bedrock, rather than by solution.

Glossary

The following terms are printed in bold-face where they first appear in a chapter. The choice of terms to be included in the glossary was predicated on the need to maintain the flow of a discussion. Many equally or more obscure technical terms are defined in the text, and can be traced through the index.

Ablation – the wastage of ice and snow by melting and other processes that reduce their mass.

Ablation till – till laid down by melting ice, usually ice that has become stagnant. It generally represents a combination of englacial and supraglacial flow tills, and these latter terms are increasingly being used today.

Accretion – deposition and accumulation of sediment.

Aggrade – to build up the floor or slope of a stream by deposition.

Albedo – the reflectivity of a surface to short-wave radiation. The albedo varies according to the colour and texture of objects. Freshly fallen snow, for example, reflects far more solar radiation back into the atmosphere than grass or forest.

Alluvium – sediment deposited by streams, and consisting largely of sand, silt, and clay.

Amorphous – material that does not have a regular arrangement of atoms.

Anticline – an upward, arch-like fold of rock.

Antidune – a small stream-bedform. Erosion occurs on the gentler downstream sides of these ridges, and deposition on the steeper upstream sides. Therefore, while the sand grains move downstream, the antidune moves upstream.

Aquifer – a layer of porous and permeable rock that allows water to easily pass through it, or a water-saturated layer such as gravel, which stores and supplies large amounts of groundwater to springs and wells.

Arc – a curved chain of islands bordering a submarine trench. Many are volcanic, owing to their association with subduction zones along plate margins. They are the result of the partial melting of subducting material, with magma rising to the surface to form volcanoes.

Argillaceous rocks – consisting largely of clay minerals. They include marl, shale, mudstone, and siltstone.

Asthenosphere – a weak layer of the Earth's mantle immediately below the lithosphere and between 100 and 240 kilometres below the surface. Because of low rock strength and poor rigidity, it is capable of being continuously deformed, thereby allowing movement of the plates above.

Backswamp – a marshy area on a floodplain outside the stream levee.

Backwash – the return flow of water down a beach from a broken wave.

Badlands – severely dissected and eroded regions with numerous deep gullies and narrow ridges.

Basal – at the base or bottom, as in basal till, which is carried or deposited at the bottom of the ice.

Basalt – dark-coloured, fine-grained, basic igneous rock formed from lava.

Batholiths – large intrusive bodies of igneous rock, usually granite, which cooled beneath a surface cover that was later removed by erosion.

Beach cusps – regularly spaced ridges of coarse sediment running perpendicular to the water line, alternating with U-shaped indentations. The horns and bays of beach cusps give a serrated or scalloped appearance to the beach at the water line.

Bedding plane – the surface that separates one bed of a sedimentary rock from another. Bedding planes represent a break in deposition, and presumably a change in the depositional environment.

Berms – the nearly horizontal terraces formed at the back of some beaches by the wave swash.

Block fields – see felsenmeer.

Boreal forest – the northern forest of coniferous trees.

Breakpoint – the point at which a wave breaks. It depends upon wave characteristics, the water depth, and the slope of the bottom.

Calving – the breakup of glacial ice into icebergs and smaller ice fragments.

Carbon-14 – see radiocarbon dating.

Cations – positively charged particles or ions.

Clastic – consisting of fragments of broken rock.

Cleavage – Cleavage planes are closely spaced planes along which a crystal or rock will tend to split. Cleavage planes in rocks can be at high angles to the bedding planes.

Cockpits – a tropical karst landform with deep hollows and conical, steep-sided hills.

Cohesive – tending to stick or hold together.

Colloidal material – very small particles suspended in water by electrical and other forces that are greater than the gravitational forces that would otherwise cause them to settle.

Colluvium – loose, weathered material brought to the foot of a cliff or some other slope by gravity (mass movement).

Compression – Compressional forces or stresses act towards each other. Compressional movements therefore involve a decrease in the length or thickness of a body, or a decrease in its volume.

Condensation – the reverse of evaporation, whereby water in the gaseous state changes into the liquid or solid state.

Conduit – a narrow underground passage that carries water.

Continental rise – the gently sloping surface (less than 1°) extending from the foot of the continental slope to the deep sea floor.

Continental shelf – gently sloping (less than 1°), shallowly submerged marginal zone of a continent. It extends from the coast out to the top of the continental slope.

Continental slope–the sloping surface at the seaward end of the continental shelf. Gradients are generally between about 2 and 5°.

Corrosion–Some workers use the term to refer to chemical weathering in general, while others restrict its use to the single process of solution.

Crescentic bar–a submarine bar consisting of a series of crescents pointing shorewards. Wavelengths often range between 200 and 500 m.

Crevasse–an open fracture in glacial ice.

Cryo–a prefix (from the Greek *kryos*, meaning cold or icy) used to refer to a variety of periglacial and glacial phenomena. Cryoplanation, for example, is the reduction in relief or the planation of an area owing to periglacial activity.

Cuesta–an asymmetrical ridge produced in gently sloping rock. The long, gentle slope conforms to the dip of the strata, while the scarp slope on the other side is shorter and usually steeper.

Debris–broken rocks and earth that have been moved to a site by gravity, ice, running water, etc.

Deflation – removal of material from a beach or other surface by wind action.

Degradation–lowering of the land surface by erosion.

Dendritic drainage–a drainage pattern characterized by irregular branching with tributaries joining the main stream at a variety of angles. Dendritic drainage develops where there is an absence of structural control.

Denudation–the washing away of surface materials to reduce irregularities, forming a surface that has a uniform level. 'Denudation' is generally used to denote all erosional processes, and is now considered to be synonymous with 'degradation'.

Deposition – the laying down of sediment that has been eroded and transported.

Deranged–Deranged surface drainage or cave systems have no pattern or form.

Diamict – a non-genetic term for any poorly sorted mud-sand-rock fragment mixture, whether, for example, glacial, periglacial, or paraglacial in origin. Tills are glacial diamicts.

Dip – 'Rock dip' refers to the inclination of the bedding planes to the horizontal.

Dip slope–a slope that conforms to the dip of the underlying rocks.

Distributary–a stream channel that splits off from the main channel and does not rejoin it.

Diurnal–pertaining to a period of 24 hours.

Drift–material laid down by glacial ice (unstratified drift or till) or by, or in, meltwater from the ice (stratified drift).

Dunes – either hills of wind-blown sand, or a type of stream bedform.

Ebb tide – the falling tide. Ebb tides are the opposite of flow tides, which are the incoming or rising tides.

Edge waves – Waves approaching a coast may generate a second set of waves, or edge waves, operating at right angles to them. Edge waves produce alternations of high and low breakers along a coast, corresponding to the position of the crests and troughs of the edge waves. Although the variations in breaker height are imperceptible, they provide a mechanism for the development of circulation cells, rip currents, and a variety of features with a regular rhythm along a coast.

Entrainment – the transportation and evacuation of debris by its incorporation or absorption into a medium such as water, air, or, especially, ice.

Entrenched – Entrenched meanders are deeply incised into their valley floor.

Erosion – the loosening or dissolving and removal of material. It includes weathering, solution, corrasion, and transportation.

Erratic – a rock fragment or boulder that has been carried by glacial ice from its place of origin and deposited in an area with a different type of bedrock.

Eustasy – a global change of sea level that is the result of a rise or fall of the ocean level, rather than a change in the level of the land. Eustatic changes in sea level in the Pleistocene were mainly the result of the growth and decay of ice sheets (glacio-eustasy), but changes in the volume of the ocean basins due to tectonics (tectono-eustasy) and sedimentation (sedimento-eustasy) were important in the Tertiary.

Exotic – introduced from different regions. Tectonic movement may result in large exotic blocks' being emplaced in zones of unrelated rock types.

Fast ice – an extensive, unbroken sheet of sea ice that is attached to the land. Fast ice is created by the *in situ* freezing of sea water.

Faults – fractures or fracture zones in which there has been movement or displacement of the rocks on either side, relative to each other.

Fauna – a broad term for animal life, sometimes used to refer to the association of animals living in some place or at some time.

Felsenmeer – a surface of loose, angular stones on the top of a mountain. Felsenmeere are generally considered to be the result of frost action.

Flora – a broad term for plant life, sometimes used to refer to the association of plants living in some place or at some time.

Fluvial – pertaining to rivers or streams. Fluvial sediments are those deposited by streams.

Gabbro – a dark, coarse-grained igneous rock.

Geomorphology – the scientific study of landforms, landscapes, and Earth surface processes.

Glacial – A glacial stage is a cold period within an ice age. During the glacial stages of the Pleistocene, there were extensive ice sheets in the northern hemisphere.

Glaciofluvial – the processes, sediments, and landforms associated with glacial meltwater streams.

Glacio-isostasy – see isostasy.

Glaciolacustrine – the processes, sediments, and landforms associated with glacial lakes.

Glaciomarine – the processes, sediments, and landforms associated with ice or meltwater streams in contact with the sea.

Gondwanaland – thought to be an ancient supercontinent existing in the southern hemisphere more than 200 million years ago, separated from the northern continent of Laurasia by the narrow Tethys Ocean. Gondwanaland was part of the even larger Precambrian supercontinent known as Pangaea. These supercontinents eventually fragmented and drifted apart, producing the land masses that exist today.

Graben – a steep-sided valley formed by a long block of land subsiding between parallel faults.

Granite – a coarse-grained, igneous, intrusive rock.

Granite gneiss – a streaky or banded metamorphic rock of granitic origin, or a primary igneous gneiss of granitic composition.

Greenhouse effect – the absorption of outgoing longwave radiation from the Earth's surface by the atmosphere. Air pollution and, particularly, the release of carbon dioxide through the combustion of fossil fuels increase atmospheric absorption. Scientists believe that this effect raises temperatures on Earth, with potentially very serious consequences.

Greenstone – basalt or some other basic volcanic rock that has been altered metamorphically.

Groundwater – the water beneath the ground surface, within saturated zones in which the hydrostatic pressure is equal or greater than atmospheric pressure.

Head – The head of a slide is the upper portion of the displaced mass along the contact with the main scarp.

Holocene – the epoch since the Pleistocene, generally taken to be since 10,000 years ago. The Pleistocene and the Holocene constitute the Quaternary, although many workers believe that the Holocene is simply an interglacial within the present ice age.

Hydraulic head – the pressure exerted by a liquid as a result of the difference in its surface level between two points.

Hydrostatic pressure – the pressure generated at any depth within a liquid at rest.

Hydroxyl – Hydroxyls consist of one hydrogen and one oxygen atom, and are known as the OH group.

Ice age – a period millions of years in length, characterized by intermittent glacial occupation of large areas of the high and middle latitudes.

Ice floes – floating, tabular blocks of ice that are thinner and flatter than icebergs. Ice floes are formed by the breakup of winter fast ice.

Icefoot – a narrow strip of intertidal ice along a coast, which does not move with the tides.

Igneous – Igneous rocks are formed by the solidification of hot, mobile material or magma from the Earth's interior.

Interfluve – the area of higher ground between two streams.

Interglacial – An interglacial stage is a period, similar to the present, of higher temperatures and ice retreat within an ice age.

Intermontane glacier – a glacier that is surrounded by mountains or mountain ranges.

Interstadial – a short period of somewhat higher temperatures and some ice retreat within a glacial stage of an ice age. Interstadials are cooler than interglacials, and ice retreat is therefore less extensive.

Interstitial – pertaining to the pores of rocks or the spaces between grains in a sediment.

Intertidal – between the high and low tidal levels on a coast. Intertidal zones are alternatively covered and uncovered by water according to the stage of the tide.

Intrusion – An intrusive body consists of igneous rocks that were forced in a molten state into the cavities in pre-existing rock strata. Intrusive ice forms from water that has been injected under pressure into sediments or rock.

Ion – an electrically charged atom (molecule).

Isostasy – a state of balance that is maintained in the crust of the Earth. Disturbance of this balance causes isostatic movements to take place, which act to restore the balance. Movements may involve uplift of the land to compensate for erosion, or depression of the land to accommodate the weight of accumulated sediment. Glacio-isostasy involves the depression of the land owing to the weight of a large body of ice, and uplift or rebound as a result of the disintegration or retreat of the ice.

Joint – a rock fracture along which there has been no appreciable movement.

Kettle – Kettle holes are enclosed depressions, often filled with water. They result from the gradual decay of blocks of glacial ice buried or partially buried in drift.

Lacustrine – pertaining to lakes.

Lag – Lag deposits consist of coarse material left behind by the removal of the finer material by, for example, waves, winds, or running water.

Latent heat – heat that is absorbed or emitted when a substance undergoes a change of state. Heat is released by condensation, freezing, and sublimation (solid to gas), and taken in, or stored, during melting, evaporation, and sublimation (gas to solid).

Lenses – Lenses are dominantly horizontal layers of ice ranging from less than a millimetre to tens of metres in thickness, and from millimetres to hundreds of metres in extent.

Lithosphere – the crust and the upper part of the mantle of the Earth's interior.

Longitudinal profile – a profile drawn down the length or long axis of a feature, such as a stream course.

Magma – molten rock intruded into the rocks in the Earth's crust, or extruded at the surface as lava.

Mass wasting – Many workers use the term synonymously with 'mass movement', although others maintain that there is a subtle difference. 'Mass wasting' is defined as the slow or rapid gravitational movement of large masses of earth material. 'Mass movement' has been defined as the movement of a portion of the land surface as a unit, as in creep or landsliding. 'Mass movement' is used in this book to refer to all gravitationally controlled movements.

Mesas – isolated, flat-topped hills with steep slopes or cliffs on at least one side.

Metamorphic – Metamorphic rocks have been altered by heat or pressure, often during mountain-building episodes.

Neoglacial – refers to periods of glacial advance in western North America in the Holocene after the warmer period between about 9,000 and 2,600 years ago (the hypsithermal).

Nunatak – an isolated rocky peak that stood out above an ice surface and may therefore have escaped glaciation.

Orogenesis – mountain building, especially by folding and thrusting.

Pack ice – floating sea ice that is not attached to the land.

Pangaea – see Gondwanaland

Passive margins – former plate boundaries that are no longer associated with significant tectonic activity. Their internal structure has subsequently been modified by subsidence and sedimentation.

Percolate – to filter or pass through fine pores or interstices, as when water percolates through porous rocks.

Perennial ice – ice that does not melt diurnally or seasonally, but persists from one year to another.

Permeability – the ease with which liquids (or gases) can pass through a rock or soil.

pH – a measure of the concentration of hydrogen ions in a solution (actually the negative of the log of the hydrogen ion concentration). The pH controls the solubility of many substances. Pure water has a neutral pH of 7, acidic solutions have a pH of less than 7, and alkaline solutions a pH greater than 7.

Piedmont glaciers – wide bodies of ice formed by the coalescence of valley glaciers at the foot of a mountain range.

Plastic–a material that can be moulded into any form, which it then retains.

Plate tectonics–the theories and concepts of continental drift, ocean floor spreading, and subduction zones.

Porosity–the ratio of the volume of the voids in a rock or soil to its total volume. Porous materials are usually also permeable, unless, as in clays, the pores are very small.

Pressure melting–the pressure-induced melting of ice near the bed.

Pressure melting point–the temperature at which a substance melts under pressure. The weight of the ice slightly decreases the melting point at depth within a glacier.

Proglacial–beyond the limits of a glacier or ice sheet. A proglacial lake is in contact with the ice front along part of its margins.

Pyroclastic–volcanic material, including lava, ash, pumice, and cinders, ejected from a volcanic vent by an explosion.

Quaternary–the most recent period of the Cenozoic era. It consists of the Pleistocene and Recent epochs, and has therefore not yet ended.

Radial flow–ice flow outwards in all directions from an ice centre, as in the spokes of a wheel.

Radiocarbon dating–one of the methods of radiometric dating. The radioactive isotope C^{14} enters living organisms from the atmosphere. Radioactive decay begins once the organism dies and is buried beneath sediments. The half life is 5,570 years, in which time half the parent C^{14} decays. The method is accurate for the dating of shells for periods up to about 20,000 years ago, and for wood and bone for periods up to 45,000 years ago, although it has been extended for periods of up to 70,000 years ago. Estimates of age are based on the original and the present radiocarbon concentration, and the constant rate of decay.

Radiometric dating–the dating of material by measuring the rate of decay of their radioactive elements. Radiocarbon and uranium-lead are among the most common applications of this technique.

Re-entrant–an indentation or recess in, for example, the margins of an ice sheet or a coastline.

Relative sea level–Changes in relative sea level refer to changes in the level of the sea relative to the land. Relative sea level could rise, for example, because of rising sea level or sinking of the land, or some combination of the two.

Relief–the difference in elevation between high and low points in an area.

Retrogression–A retrogressive slide or flow enlarges or retreats in the opposite direction to the direction of movement of the material.

Rillwash–the removal of surface material by the flow of water in numerous tiny channels.

Rupture surface–in mass movements, the surface along which rock or soil moves.

Saprolite – deeply weathered bedrock that has retained its coherency.

Scarp – The main scarp of a slide is the steep slope around its periphery.

Scarp slope – a steep slope in dipping rocks.

Sediment – solid material that has been transported and deposited by air, water, or ice.

Sedimentary – Sedimentary rock is formed from the accumulation, in a layered sequence, of material derived from pre-existing rocks or from organic sources.

Segregated ice – ice lenses or bands of clear ice derived from the freezing of soil water, or more particularly, by the freezing of groundwater drawn upwards by capillary action.

Shear – Shearing involves two adjacent parts of a solid sliding past each other along a shared surface (shear fault).

Sheetwash – the removal of surface material on gentle slopes by the shallow, unchannelled flow of water.

Shingle – stony beach material that is coarser than gravel.

Sills – essentially horizontal sheets of intrusive material (usually diorite) that were injected along the bedding planes of crustal rocks.

Skin-type deformation – occurs when the upper part of the Earth's crust is deformed and detached from the lower part.

Slopewash – the transport of surface material downslope by running water.

Spatial – pertaining to position. 'Spatial variations' refers to differences in some phenomenon between different sites.

Stadial – a substage in a glacial stage when glaciers temporarily advance or became stationary.

Stillstand – a period when relative sea level or the position of ice margins remain constant.

Stoss slope – the slope that faces the direction from which the ice came, as in the steeper stoss slope of a drumlin.

Stratified – in distinct beds. Sedimentary and water-laid sediments, for example, including glaciofluvial, are laid down in beds.

Stream – a flowing, channelized body of water of any size. The term is generally used by geomorphologists in preference to 'river'.

Stress – the internal forces generated within a body by external forces. Stress is compressive if the external forces are applied towards each other; tensional if they act away from each other; and tangential or shearing if they act tangentially to each other.

Strike – The strike of a rock is the direction of a horizontal line drawn across the plane of the bed at right angles to the dip.

Subaerial – features or processes on the Earth's surface, rather than below the surface or the sea.

Sublimation – transformation of a solid directly to gas, or vice versa, without an intermediate liquid stage.

Subtidal – below the low tidal level on coasts. Subtidal zones are under water at all stages of the tide.

Successor basins – deep, subsiding troughs that overlie the eroded rocks of an orogenic belt. They develop as the crust in the mobile belt becomes increasingly continentalized.

Supratidal – above the high tidal level on a coast. Supertidal zones are above the water at all stages of the tide.

Surf zone – the area of the nearshore zone extending from the breakers to the swash zone.

Swash – the turbulent water that moves up a beach after a wave has broken.

Swash zone – the area of the nearshore zone where the beach is alternatively covered and uncovered by the incoming swash and the outgoing backwash.

Synclines – downfolds or basin-shaped folds in rocks.

Tectonic – pertaining to the internal forces that act to deform or uplift the crust of the Earth.

Tectonically active areas – areas where the crust is being deformed or uplifted.

Temporal – pertaining to time. 'Temporal variations' refers to changes in some phenomenon through time.

Tension – stresses acting against each other so that they pull solids apart.

Thermo erosion – erosion as a result of melting. This usually involves the melting and subsequent erosion of ice-rich sediments or ground ice along the banks of rivers, lakes, or seas in the Arctic.

Thrust faults – reverse faults with a long angle of dip to the horizontal.

Thrusting – shearing movement in a glacier along a slip plane or fault, especially in the thin ice near the glacier margins.

Till – unsorted and unstratified material carried and laid down by glacial ice.

Toe – The toe of a slide is the edge of the displaced material furthest away from the main scarp.

Tor – a small hill or stack of well-jointed rock rising abruptly from a slope or hilltop.

Transform fault – a massive example of a fault that is transverse to the strike of the folded rocks. They end abruptly where they are transformed into other types of structure.

Transgression – progressive submergence of the land through a rise in sea level or a drop in the level of the land.

Transpiration – the process by which water is taken up by the roots of vegetation, passed into the leaves, and then returned to the atmosphere.

Tundra – a vegetational zone of lichen, mosses, sedges, and dwarf trees in high latitudes.

Turbidity currents – rapidly flowing submarine currents consisting of dense mixtures of water and sediment. They flow downslope under gravity, carrying large amounts of sediment onto the deep sea floor.

Unconformable – An unconformity is the contact between older and younger rocks, representing a period of erosion or non-deposition of sediment. It therefore marks a break in the geological record or a period of unrecorded time.

Unconsolidated – loose material as opposed to rock.

Viscous – adhesive or sticky. A viscous liquid does not immediately deform under stress.

Water table – the level in the ground below which all pore spaces are saturated with water.

Wave period – the time it takes a wave to travel a distance equal to one wavelength. It can be measured by timing the passage of two wave crests past a fixed point.

Wisconsin glacial – the last glacial stage in North America.

Zone of accumulation – the lower zone on a slide, where enough displaced material accumulates to rise above the original ground surface.

Zone of depletion – the upper zone on a slide, where the displaced material lies below the original ground surface. It is essentially the crater created by a slide.

The S.I. System of Measurement

The international system of units (S.I.) used in this book is based upon six primary units:

	UNIT	SYMBOL
length	metre	m
mass	kilogram(me)	kg
time	second	s
electrical current	ampere	A
temperature	degree Kelvin	°K
luminous intensity	candela	cd

Other units are derived through combinations of symbols, such as km/s or km s^{-1}, m^2 (square metres), and m^3 (cubic metres). The use of seconds in geological work involving hundreds if not thousands of years would, of course, be unworkable. There is therefore a growing tendency to use the symbol 'a' (annus) for a year; hence the use of ka for thousands of years.

Multiples and sub-multiples are confined to stages of 10^3; that is, steps one thousand greater or one thousand less (few exceptions, such as the centimetre, which is 10^{-2}m or one hundredth less, and the hectare, an area equal to 10,000 square metres, are allowed). The metric ton is 1,000 kg or 1 Mg (megagram).

MULTIPLYING FACTOR	PREFIX	SYMBOL
10^{+12}	tera	T
10^{+9}	giga	G
10^{+6}	mega	M
10^{+3}	kilo	k
10^{-3}	milli	m
10^{-6}	micro	u
10^{-9}	nano	n
10^{-12}	pico	p
10^{-15}	femto	f
10^{-18}	atto	a

For example, the metre is the standard unit of length. A *kilo*metre (*km*) is therefore 10^3 metres (i.e., $10 \times 10 \times 10$ metres) or a thousand metres. Similarly a *u*m, a *micro*metre, is 10^{-6} ($10 \times 10 \times 10 \times 10 \times 10 \times 10$th of a metre), or a millionth of a metre. A km^3 is a cubic kilometre.

Some rules for writing S.I. units
- Abbreviations do not take an 's' in the plural; hence 2 km, not 2 kms.
- Symbol abbreviations are not followed by a period (unless at an end of a sentence, etc.).
- Multiplying prefixes (m, k, G, T, etc.) are put immediately next to the unit; hence km, not k m.
- SI symbols and numerals should be separated from each other by a space; hence 14 km.

Conversion factors

1 inch = 25.40 mm	1 mm = 0.0394 inches
1 foot = 0.3048 m	1 m = 3.281 feet
1 mile = 1.609 km	1 km = 0.6214 miles

Area and Volume

There are 10 mm in a cm, 100 cm in a metre, and 1,000 m in a km. However, there are 100×100 cm^2 in a m^2, and $1,000 \times 1,000 \times 1,000$ m^3 in a km^3.

References

The references given in this book are heavily biased towards recent publications. There is, of course, a wealth of valuable information in the older literature, as well as in many newer publications, which could not be listed below. Many of these publications can be traced through the references given in the following books and articles.

Adams, J., and Basham, P. 1989: The seismicity and seismotectonics of Canada east of the Cordillera. *Geoscience Canada* 16, pp. 3-16.

Allen, C.C., Jercinovic, M.J., and Allen, J.S.B. 1982: Subglacial volcanism in north-central British Columbia. *Journal of Geology* 90, pp. 699-715.

Anderson, L.W. 1978: Cirque glacier erosion rates and characteristics of neoglacial tills, Pangnirtung Fiord area, Baffin Island, NWT Canada. *Arctic and Alpine Research* 10, pp. 749-60.

Andrews, J.T., 1982: Comment on 'New evidence from beneath the western North Atlantic for the depth of glacial erosion in Greenland and North America' by E.P. Laine. *Quaternary Research* 17, pp. 123-4.

Andrews, J.T., ed. 1985: *Quaternary Environments* (Allen and Unwin, London).

Andrews, J.T. 1987: The late Wisconsin glaciation and deglaciation of the Laurentide ice sheet. In: W.F. Ruddiman and H.E. Wright (eds), *North America and Adjacent Oceans During the Last Deglaciation*, Geological Society of America, The Geology of North America, K-3, pp. 13-37.

Andrews, J.T., and Miller, G.H. 1984: Quaternary glacial and nonglacial correlations for the eastern Canadian Arctic. In: R.J. Fulton (ed.), *Quaternary Stratigraphy of Canada – a Canadian Contribution to IGCP Project 24*. Geological Survey of Canada, Paper 84-10, pp. 101-16.

Andrews, J.T., Shilts, W.W., and Miller, G.H. 1983: Multiple deglaciation of the Hudson Bay Lowlands, Canada, since deposition of the Missinaibi (last-interglacial?) Formation. *Quaternary Research* 19, pp. 18-37.

Aylsworth, J.M., and Shilts, W.W. 1989: Glacial features around the Keewatin ice divide: Districts of Mackenzie and Keewatin. *Geological Survey of Canada*, Paper 88-24.

Banerjee, I., and McDonald, B.C. 1975: Nature of esker sedimentation. In: A.V. Jopling and B.C. McDonald (eds), *Glaciofluvial and Glaciolacustrine Sedimentation*, Society Economic Palaeontologists and Mineralogists, Special Publication 23, pp. 132-54.

Barendregt, R.W., and Ongley, E.D. 1977: Piping in the Milk River Canyon, southeastern Alberta – a contemporary dryland geomorphic process. *International Association of Hydrological Sciences*, Publication 122, pp. 233-43.

Bastin, R.M., and Mathews, W.H. 1979: Selective weathering of granitic clasts. *Canadian Journal of Earth Sciences* 16, pp. 215-22.

Beaty, C.B. 1975: Coulee alignment and the wind in southern Alberta, Canada. *Geological Society of America Bulletin* 86, pp. 119-28.

Béland, J. 1956: Nicolet Landslide. *Geological Association of Canada Proceedings* 8, pp. 143-56.

Bell, M., and Laine, E.P. 1985: Erosion of the Laurentide Region of North America by glacial and glaciofluvial processes. *Quaternary Research* 23, pp. 154-74.

Bergeron, N., and Roy, A.G. 1985: Le rôle de la végétation sur la morphologie d'un petit cours d'eau. *Géographie Physique et Quaternaire* 39, pp. 323-6.

Bernard, C. 1971: Les marques sous-glaciaires d'aspect plastique sur la roche en place (p-forms). *Revue de Géographie de Montréal* 25, pp. 111-27 and 265-76.

Bird, E.C.F., and Schwartz, M.L., eds. 1985: *The World's Coastline* (Van Nostrand Reinhold, New York). See: British Columbia by Owens, E.H., and Harper, J.R. pp. 11-14; Atlantic Canada by McCann, S.B. pp. 235-9; Arctic Canada by Bird, J.B. pp. 241-51; and Great Lakes by Carter, C.H., and Haras, W.S. pp. 253-60.

Bird, J.B. 1963: Limestone terrains in southern Arctic Canada. *Proc. Ist Permafrost International Conference*, pp. 115-21.

Bird, J.B. 1967: *The Physiography of Arctic Canada* (Johns Hopkins Press, Baltimore).

Bird, J.B. 1972: *The Natural Landscapes of Canada* (Wiley, Toronto).

Bostock, H.S. 1970: Physiographic subdivisions of Canada. In: R.J.W. Douglas (ed.), *Geology and Economic Minerals of Canada*, Geological Survey of Canada, Economic Geology Report 1, pp. 10-30.

Bouchard, M. 1985: Weathering and weathering residuals on the Canadian Shield. *Fennia* 163, pp. 327-32.

Bouchard, M., and Godard, A. 1984: Les altérites du Bouclier Canadien: premier bilan d'une campagne de reconnaissance. *Géographie Physique et Quaternaire* 38, pp. 149-63.

Boulton, G.S. 1974: Processes and patterns of glacial erosion. In: D.R. Coates (ed.), *Glacial Geomorphology* (State University of New York, Binghamton, N.Y.), pp. 41-87.

Boulton, G.S., Smith, G.D., Jones, A.S., and Newsome, J. 1985: Glacial geology and glaciology of the last mid-latitude ice sheets. *Journal of the Geological Society of London* 142, pp. 447-74.

Bovis, M.J. 1982: Uphill-facing (antislope) scarps in the Coast Mountains, southwest British Columbia. *Geological Society of America Bulletin* 93, pp. 804-12.

Bovis, M.J. 1985: Earthflows in the interior plateau, southwest British Columbia. *Canadian Geotechnical Journal* 22, pp. 313-34.

Brook, G.A., and Ford, D.C. 1974: Nahanni karst: unique northern landscape. *Canadian Geographical Journal* 88, pp. 36-43.

Brook, G.A., and Ford, D.C. 1978: The origin of labyrinth and tower karst and the climatic conditions necessary for their development. *Nature* 275, pp. 493-6.

Brown, R.J.E. 1970: *Permafrost in Canada* (University of Toronto Press, Toronto).

Bryant, E.A. 1983: Sediment characteristics of some Nova Scotian beaches. *Maritime Sediments and Atlantic Geology* 19, pp. 127-42.

Butler, D.R. 1989: Canadian landform examples – 11 Subalpine snow avalanche slopes. *Canadian Geographer* 33, pp. 269-73.

Campbell, I.A. 1987: Canadian landform examples – 3 Badlands of Dinosaur Provincial Park. *Canadian Geographer* 31, pp. 82-7.

Carter, R.W.G., Forbes, D.L., Jennings, S.C., Orford, J.D., Shaw, J., and Taylor, R.B. 1989: Barrier and lagoon coast evolution under differing relative sea-level regimes: examples from Ireland and Nova Scotia. *Marine Geology* 88, pp. 221-42.

Chapman, L.J., and Putnam, D.F. 1966: *The Physiography of Southern Ontario* (University of Toronto Press, Toronto), 2nd ed.

Chappell, J. 1983: A revised sea-level record for the last 300,000 years from Papua New Guinea. *Search* 14, pp. 99-101.

Christiansen, E.A. 1967: Preglacial valleys in southern Saskatchewan. Saskatchewan Research Council (Geology Division), Map no. 3.

Church, M. 1972: Baffin Island sandurs: a study of Arctic fluvial processes. *Canadian Geological Survey*, Bulletin 216.

Church, M., and Ryder, J.M. 1972: Paraglacial sedimentation: a consideration of fluvial processes conditioned by glaciation. *Geological Society of America Bulletin* 83, pp. 3059-72.

Church, M., and Slaymaker, O. 1989: Disequilibrium of Holocene sediment yield in glaciated British Columbia. *Nature* 337, pp. 452-4.

Church, M., Stock, R.F., and Ryder, J.M. 1979: Contemporary sedimentary environments on Baffin Island, NWT., Canada: debris slope accumulations. *Arctic and Alpine Research* 11, pp. 371-402.

Cinq-Mars J., and Lauriol, B. 1985: Le karst de Tsi-It-Toh-Choh: notes préliminaires sur quelques phénomènes karstiques du Yukon septentrional, Canada. *Annales de la Géologique de Belgique* 108, pp. 185-95.

Clague, J.J. 1987: Canadian landform examples – 5 Rock avalanches. *Canadian Geographer* 31, pp. 278-82.

Clague, J.J. 1988: Quaternary stratigraphy and history, Quesnel, British Columbia. *Géographie Physique et Quaternaire* 42, pp. 279-88.

Clague, J.J., and Bornhold, B.D. 1980: Morphology and littoral processes of the Pacific coast of Canada. In: S.B. McCann (ed.), *The Coastline of Canada*, Geological Survey of Canada, Paper 80-10, pp. 339-80.

Clague, J.J., Luternauer, J.L., and Hebda, R.J. 1983: Sedimentary environments and postglacial history of the Fraser Delta and lower Fraser Valley, British Columbia. *Canadian Journal of Earth Sciences* 20, pp. 1314-26.

Clague, J.J., and Souther, J.G. 1982: The Dusty Creek landslide on Mount Cayley, British Columbia. *Canadian Journal of Earth Sciences* 19, pp. 524-39.

Clark, P.U. 1988: Glacial geology of the Torngat Mountains, Labrador. *Canadian Journal of Earth Sciences* 25, pp. 1184-98.

Clark, T.H., and Stearn, C.W. 1960: *The Geological Evolution of North America* (Ronald Press, New York).

Clement, B., Landry, B., and Yergeau, M. 1976: Déchaussement postglaciaire de filons de quartz dans les Appalaches Québecoises. *Geografiska Annaler* 58A, pp. 111-14.

Clément, P., and De Kimpe, C.R. 1977: Geomorphological conditions of gabbro weathering at Mount Megantic, Québec. *Canadian Journal of Earth Sciences* 14, pp. 2262-73.

Clément, P., and Poulin, A. 1975: La fossilisation des réseaux de vallées aux environs de Sherbrooke, Québec. *Revue de Géographie de Montréal* 29, pp. 167-71.

Cowell, D.W., and Ford, D.C. 1975: The Wodehouse Creek karst, Grey County, Ontario. *Canadian Geographer* 19, pp. 196-205.

Crozier, M.J. 1975: On the origin of the Peterborough drumlin field: testing the dilatancy theory. *Canadian Geographer* 19, pp. 181-95.

Cruden, D.M. 1976: Major rock slides in the Rockies. *Canadian Geotechnical Journal* 13, pp. 8-20.

Cruden, D.M. 1985: Rock slope movements in the Canadian Cordillera. *Canadian Geotechnical Journal* 22, pp. 528-40.

Cruden, D.M., Bornhold, B.D., Chagnon, J.-Y., Evans, S.G., Heginbottom, J.A., Locat, J., Moran, K., Piper, D.J.W., Powell, R., Prior, D., Quigley, R.M., and Thomson, S. 1989: Landslides: extent and economic significance in Canada. International Geological Congress Washington, July 1989, pp. 1-23.

David, P.P. 1977: Sand dune occurrences of Canada. Indian and Northern Affairs, National Parks Branch, Contract no. 74-230 Report.

David, P.P. 1981: Stabilized dune ridges in northern Saskatchewan. *Canadian Journal of Earth Sciences* 18, pp. 286-310.

Davidson-Arnott, R.G.D., and Pember, G.F. 1980: Morphology and sedimentology of multiple bar systems, southern Georgian Bay, Ontario. In: S.B. McCann (ed.), *The Coastline of Canada*, Geological Survey of Canada, Paper 80-10, pp. 417-28.

Desloges, J.R., and Church, M. 1989: Canadian landform examples – 13 Wandering gravel-bed rivers. *Canadian Geographer* 33, pp. 360-4.

Dickinson, W.T., Scott, A., and Wall, G. 1975: Fluvial sedimentation in southern Ontario. *Canadian Journal of Earth Sciences* 12, pp. 1813-19.

Dionne, J.-C. 1978: Le glaciel en Jamésie et en Hudsonie, Québec subarctique. *Géographie Physique et Quaternaire* 32, pp. 3-70.

Dionne, J.-C. 1987: Tadpole rock (rocdrumlin): a glacial streamline moulded form. In: J. Menzies and J. Rose (eds), *Drumlin Symposium* (A.A. Balkema, Rotterdam), pp. 149-59.

Dionne, J.-C. 1988: Characteristic features of modern tidal flats in cold regions. In: P.L. de Boer (ed.), *Tide-influenced Sedimentary Environments and Facies* (D. Reidel, Dordrecht, The Netherlands), pp. 301-32.

Dionne, J.-C., and Michaud, Y. 1986: Note sur l'altération chimique en milieu périglaciaire, Hudsonie, Québec subarctique. *Revue de Géomorphologie Dynamique* 35, pp. 81-92.

Dohler, G. 1966: *Tides in Canadian Waters*. Canadian Hydrographic Service, Marine Sciences Branch, Department of Mines and Technical Surveys, Ottawa.

Dowdeswell, E.K., and Andrews, J.T. 1985: The fiords of Baffin Island: description and classification. In: J.T. Andrews (ed.), *Quaternary Environments* (Allen and Unwin, London), pp. 93-123.

Drake, J.J., and Ford, D.C. 1974: Solutional erosion in the southern Canadian Rockies. *Canadian Geographer* 20, pp. 158-70.

Dredge, L.A., Nixon, F.M., and Richardson, R.J. 1986: Quaternary geology and geomorphology of northwestern Manitoba. *Geological Survey of Canada*, Memoir 418.

Dredge, L.A., and Thorleifson, L.H. 1987: The middle Wisconsinan history of the Laurentide ice sheet. *Géographie Physique et Quaternaire* 41, pp. 215-35.

Dubois, J.-M.M., and Dionne, J.-C. 1985: The Québec North Shore Moraine System: a major feature of late Wisconsin deglaciation. *Geological Society of America*, Special Paper 197, pp. 125-33.

Dury, G.H. 1964: Principles of underfit streams. *U.S. Geological Survey*, Professional Paper 452-A.

Dyke, A.S., Dredge, L.A., and Vincent, J.-S. 1982: Configuration and dynamics of the Laurentide ice sheet during the late Wisconsin maximum. *Géographie Physique et Quaternaire* 36, pp. 5-14.

Dyke, A.S., and Morris, T.F. 1988: Canadian landform example-7 Drumlin fields, dispersal trains, and ice streams in Arctic Canada. *Canadian Geographer* 32, pp. 86-90.

Dyke, A.S., and Prest, V.K. 1987: Late Wisconsinan and Holocene history of the Laurentide ice sheet. *Géographie Physique et Quaternaire* 41, pp. 237-63.

Eisbacher, G.H. 1979a: First-order regionalization of landslide characteristics in the Canadian Cordillera. *Geoscience Canada* 6, pp. 69-79.

Eisbacher, G.H. 1979b: Cliff collapse and rock avalanches (sturzstroms) in the Mackenzie Mountains, northwestern Canada. *Canadian Geotechnical Journal* 16, pp. 309-34.

Eisbacher, G.H., and Clague, J.J. 1984: Destructive mass movements in high mountains: hazard and management. *Geological Survey of Canada*, Paper 84-16.

England, J. 1976: Late Quaternary glaciation of the eastern Queen Elizabeth Islands, Northwest Territories, Canada. *Quaternary Research* 6, pp. 185-202.

Evans, S.G. 1982: Landslides and superficial deposits in urban areas of British Columbia. *Canadian Geotechnical Journal* 19, pp. 269-88.

Evans, S.G., and Clague, J.J. 1989: Rain-induced landslides in the Canadian Cordillera, July 1988. *Geoscience Canada* 16, pp. 193-200.

Evans, S.G., Clague, J.J., Woodsworth, G.J., and Hungr, O. 1989: The Pandemonium Creek rock avalanche, British Columbia. *Canadian Geotechnical Journal* 26, pp. 427-46.

Eyles, N., and Clark, B.M. 1988: Last interglacial sediments of the Don Valley Brickyard, Toronto, Canada, and their paleoenvironmental significance. *Canadian Journal of Earth Sciences* 25, pp. 1108-22.

Fahey, B.D., and LeFebure, T.H. 1988: The freeze-thaw weathering regime at a section of the Niagara Escarpment on the Bruce Peninsula, southern Ontario, Canada. *Earth Surface Processes and Landforms* 13, pp. 293-304.

Falconer, G., Ives, J.D., Lèken, O.H., and Andrews, J.T. 1965: Major end moraines in eastern and central Arctic Canada. *Geographical Bulletin* 7, pp. 137-53.

Farvolden, R.N. 1963: Bedrock channels of southern Alberta. In: Early contributions to the groundwater hydrology of Alberta. *Research Council of Alberta*, Bulletin 12, pp. 63-75.

Fisher, D.A., Reeh, N., and Langley, K. 1985: Objective reconstructions of the late Wisconsinan Laurentide ice sheet and the significance of deformable beds. *Géographie Physique et Quaternaire* 39, pp. 229-38.

FitzGibbon, J.E. 1981: Thawing of seasonally frozen ground in organic terrain in central Saskatchewan. *Canadian Journal of Earth Sciences* 18, pp. 1492-6.

Flint, R.F. 1943: Growth of the North American ice sheet during the Wisconsin age. *Geological Society of America Bulletin* 54, pp. 325-62.

Flint, R.F. 1971: *Glacial and Quaternary Geology* (John Wiley, New York).

Ford, D.C. 1967: Sinking creeks of Mt Tupper a remarkable groundwater system in Glacier National Park, BC. *Canadian Geographer* 11, pp. 49-52.

Ford, D.C. 1979: A review of alpine karst in the southern Rocky Mountains of Canada. *National Speleological Society*, Bulletin 41, pp. 53-65.

Ford, D.C. 1983a: Effects of glaciations upon karst aquifers in Canada. *Journal of Hydrology* 61, pp. 149-58.

Ford, D.C. 1983b: Karstic interpretation of the Winnipeg aquifer. *Journal of Hydrology* 61, pp. 177-80.

Ford, D.C. 1983c: Alpine karst systems at Crowsnest Pass, Alberta–British Columbia. *Journal of Hydrology* 61, pp. 187-92.

Ford, D.C. ed. 1983d: Castleguard cave and karst, Columbia Icefields area, Rocky Mountains of Canada: a symposium. *Arctic and Alpine Research* 15, pp. 425-554.

Ford, D.C. 1987: Effects of glaciations and permafrost upon the development of karst in Canada. *Earth Surface Processes and Landforms* 12, pp. 507-21.

Ford, D.C., and Ewers, R.O. 1978: The development of limestone cave systems in the dimensions of length and depth. *Canadian Journal of Earth Sciences* 15, pp. 1783-98.

Ford, D.C., Schwarcz, H.P., Drake, J.J., Gascoyne, M., Harmon, R.S., and Latham, A.G. 1981: Estimates of the existing relief within the southern Rocky Mountains of Canada. *Arctic and Alpine Research* 13, pp. 1-10.

Ford, D.C., and Williams, P.W. 1989: *Karst Geomorphology and Hydrology* (Unwin Hyman, London).

Fraser, J.K. 1959: Freeze-thaw frequencies and mechanical weathering in Canada. *Arctic* 12, pp. 40-53.

French, H.M. 1976: *The Periglacial Environment* (Longman, London).

French, H.M., Bennett, L., and Hayley, D.W. 1986: Ground ice conditions near Rea Point and on Sabine Peninsula, eastern Melville Island. *Canadian Journal of Earth Sciences* 23, pp. 1389-1400.

Fulton, R.J. ed. 1984: A Canadian contribution to IGCP Project 24, *Geological Survey of Canada*, Paper 84-10.

Fulton, R.J. ed. 1989: *Quaternary Geology of Canada and Greenland*. Geological Survey of Canada – Geological Society of America, The Geology of North America, K-1.

Gabrielse, H., and Yorath, C.J. 1989: DNAG #4. The Cordilleran Orogen in Canada. *Geoscience Canada* 16, pp. 67-83.

Gadd, N.R. 1964: Moraines in the Appalachian region of Quebec. *Geological Society of America Bulletin* 75, pp. 1249-54.

Gadd, N.R., ed. 1989: The late Quaternary development of the Champlain sea basin. *Geological Association of Canada*, Special Paper 35.

Gagnon, H. 1972: La photo aérienne dans les études de glissement de terrain. *Revue de Géographie de Montréal* 26, pp. 381-406.

Gardner, J.S. 1983: Rockfall frequency and distribution in the Highwood Pass area, Canadian Rocky Mountains. *Zeitschrift fur Geomorphologie* 27, pp. 311-24.

Geological Society of America 1952: Pleistocene eolian deposits of the United States, Alaska and parts of Canada – map, scale 1:2,500,000.

Gerath, R.F., and Hungr, O. 1983: Landslide terrain, Scatter River Valley, northeastern British Columbia. *Geoscience Canada* 10, pp. 30-2.

Good, T.R., and Bryant, I.D. 1985: Fluvio-aeolian sedimentation, an example from Banks Island, NWT, Canada. *Geografiska Annaler* 67A, pp. 33-46.

Grant, D.R. 1977: Glacial style and ice limits, the Quaternary stratigraphic record, and changes of land and ocean level in the Atlantic Provinces, Canada. *Géographie Physique et Quaternaire* 31, pp. 247-60.

Grant, D.R. 1980: Quaternary sea-level change in Atlantic Canada as an indication of crustal delevelling. In: N.-A. Morner (ed.), *Earth Rheology, Isostasy and Eustasy* (John Wiley, Chichester), pp. 201-14.

Gravenor, C.P. 1975: Erosion by continental ice sheets. *American Journal of Science* 275, pp. 594-604.

Greenwood, B., and Davidson-Arnott, R.G.D. 1975: Marine bars and nearshore sedimentary processes, Kouchibouguac Bay, New Brunswick. In: J.R. Hails and A. Carr (eds), *Nearshore Sediment Dynamics and Sedimentation* (John Wiley, New York), pp. 123-50.

Grieve, R.A.F., and Robertson, P.B. 1987: Terrestrial impact structures – map. Geological Society of Canada Map 1658A (included as a supplement in *Episodes* 10 [2] p. 86).

Gwyn, Q.H.J., and Cowan, W.R. 1978: The origin of the Oak Ridges and Orangeville Moraines of southern Ontario. *Canadian Geographer* 22, pp. 345-52.

Hale, P.B., and McCann, S.B. 1982: Rhythmic topography in a mesotidal low-wave-energy environment. *Journal of Sedimentary Petrology* 52, pp. 415-30.

Hamelin, L.-E., and Cook, F.A. 1967: *Illustrated Glossary of Periglacial Phenomena* (University of Laval Press, Québec).

Hare, F.K., and Thomas, M.K. 1974: *Climate Canada* (Wiley, Toronto).

Harland, W.B., Cox, A.V., LLewellyn, P.G., Pickton, C.A.G., Smith, A.G., and Walters, R. 1982: *A Geologic Time Scale* (Cambridge University Press, Cambridge).

Harris, S.A. 1981: Distribution of zonal permafrost landforms with freezing and thawing indices. *Erdkunde* 35, no. 2, pp. 81-90.

Harris, S.A. 1986: Permafrost distribution, zonation and stability along the eastern ranges of the Cordillera of North America. *Arctic* 39, pp. 29-38.

Harris, S.A., and Gustafson, C.A. 1988: Retrogressive

slumps, debris flows and river valley development in icy, unconsolidated sediments on hills and mountains. *Zeitschrift fur Geomorphologie* 32, pp. 441-55.

Harrison, V.F., Gow, W.A., and Ivarson, K.C. 1966: Leaching of uranium from Eliot Lake ore in the presence of bacteria. *Canadian Mining Journal* 87, pp. 64-7.

Harry, D.G., French, H.M., and Pollard, W.H. 1988: Massive ground ice and ice-cored terrain near Sabine Point, Yukon Coastal Plain. *Canadian Journal of Earth Sciences* 25, pp. 1846-56.

Heginbottom, J.A., and eleven others, in press: The extent and economic significance of landsliding in Canada. *Geological Survey of Canada*, Paper.

Hickin, E.J. 1974: The development of meanders in natural river-channels. *American Journal of Science* 274, pp. 414-42.

Hickin, E.J. 1984: Vegetation and river channel dynamics. *Canadian Geographer* 28, pp. 111-26.

Hickin, E.J. 1986: Concave-bank benches in the floodplains of Muskwa and Fort Nelson Rivers, British Columbia. *Canadian Geographer* 30, pp. 111-22.

Hickin, E.J., and Nanson, G.C. 1975: The character of channel migration on the Beatton River, northeast British Columbia, Canada. *Geological Society of America Bulletin* 86, pp. 487-94.

Hicock, S.R., Kristjansson, F.J., and Sharpe, D.R. 1989: Carbonate till as a soft bed for Pleistocene ice streams on the Canadian Shield north of Lake Superior. *Canadian Journal of Earth Sciences* 26, pp. 2249-54.

Hillaire-Marcel, C., Grant, D.R., and Vincent, J.-S. 1980: Comment and reply on "Keewatin ice sheet-re-evaluation of the traditional concept of the Laurentide ice sheet" and "Glacial erosion and ice sheet divides, northeastern Laurentide ice sheet, on the basis of the distribution of limestone erratics". *Geology* 8, pp. 466-8.

Hillaire-Marcel C., and Occhietti, S. 1980: Chronology, paleogeography and paleoclimatic significance of the late and post-glacial events in eastern Canada. *Zeitschrift fur Geomorphologie* 24, pp. 373-92.

Hillaire-Marcel, C., Occhietti, S., and Vincent, J.-S. 1981: Sakami Moraine, Quebec: a 500-km-long moraine without climatic control. *Geology* 9, pp. 210-14.

Hudec, P.P. 1973: Weathering of rocks in Arctic and sub-arctic environment. In: J.D. Aitken and D.J. Glass (eds), *Canadian Arctic Geology*, Geological Association of Canada Symposium, pp. 313-35.

Huntley, D.A. 1980: Edge waves in a crescentic bar system. In: S.B. McCann (ed.), *The Coastline of Canada*, Geological Survey of Canada, Paper 80-10, pp. 111-21.

Hydrological Atlas of Canada 1978: Energy, Mines and Resources, Cartographic Information and Distribution Centre, Ottawa, Canada.

Jackson, L.E. Jr, MacDonald, G.M., and Wilson, M.C. 1982: Paraglacial origin for terraced river sediments in Bow Valley, Alberta. *Canadian Journal of Earth Sciences* 19, pp. 2219-31.

John, B.S., ed. 1979: *The Winters of the World* (David and Charles, Newton Abbot, Devon), pp. 29-57.

Johnson, D.W. 1925: *The New England-Acadian Shoreline* (John Wiley, New York).

Johnson, P.G. 1978: Rock glacier types and their drainage systems, Grizzly Creek, Yukon Territory. *Canadian Journal of Earth Sciences* 15, pp. 1496-1507.

Johnson, P.G. 1984a: Paraglacial conditions of instability and mass movement. A discussion. *Zeitschrift fur Geomorphologie* 28, pp. 235-50.

Johnson, P.G. 1984b: Rock glacier formation by high-magnitude low-frequency slope processes in the southwest Yukon. *Annals of the Association of American Geographers* 74, pp. 408-19.

Jones, N. 1982: The formation of glacial flutings in east-central Alberta. In: R. Davidson-Arnott, W. Nickling, and B.D. Fahey (eds), *Glacial, Glaciofluvial and Glaciolacustrine Systems* (Geo Books, Norwich), 6th Guelph Symposium in Geomorphology, pp. 49-70.

Karolyi, M.S., and Ford, D.C. 1983: The Goose Arm karst, Newfoundland, Canada. *Journal of Hydrology* 61, pp. 181-5.

Karrow, P.F., and Calkin, P.E., eds. 1985: Quaternary evolution of the Great Lakes, *Geological Association of Canada*, Special Paper 30.

Kehew, A.E., and Lord, M.L. 1989: Canadian landform examples-12 Glacial lake spillways of the central Interior Plains, Canada-USA. *Canadian Geographer* 33, pp. 274-7.

Kennedy, B.A., and Melton, M.A. 1972: Valley asymmetry and slope forms in a permafrost area in the Northwest Territories, Canada. In: R.J. Price and D.E. Sugden (eds), *Polar Geomorphology*, Institute of British Geographers, Special Publication 4, pp. 107-21.

Klassen, R.W. 1972: Wisconsin events and the Assiniboine and Qu'Appelle Valleys of Manitoba and Saskatchewan. *Canadian Journal of Earth Sciences* 9, pp. 544-60.

Knight, R.J. 1980: Linear sand bar development and tidal current flow in Cobequid Bay, Bay of Fundy, Nova Scotia. In: S.B. McCann (ed.), *The Coastline of Canada*, Geological Survey of Canada, Paper 80-10, pp. 123-52.

Knight, R.J., and Dalrymple, R.W. 1976: Winter conditions in a macrotidal environment, Cobequid Bay, Nova Scotia. *Revue de Géographie de Montréal* 30, pp. 65-86.

Knighton, A.D. 1976: Stream adjustment in a small Rocky Mountain basin. *Arctic and Alpine Research* 8, pp. 197-212.

Kor, P.S.G., and Teller, J.T. 1986: Canadian landform examples–1 Ouimet Canyon, Ontario–deep erosion by glacial meltwater. *Canadian Geographer* 30, pp. 273-6.

Kostaschuk, R.A. 1987: Canadian landform examples – 4 Some British Columbian fjord deltas. *Canadian Geographer* 31, pp. 180-4.

Kostaschuk, R.A., MacDonald, G.M., and Jackson, L.E. Jr. 1987: Canadian landform examples–6 Rocky Mountain alluvial fans. *Canadian Geographer* 31, pp. 366-8.

Kostaschuk, R.A., MacDonald, G.M., and Putnam, P.E. 1986: Depositional processes and alluvial fan-drainage basin morphometric relationships near Banff, Alberta, Canada. *Earth Surface Processes and Landforms* 11, pp. 471-84.

Lachenbruch, A.H. 1962: Mechanics of thermal contraction cracks and ice-wedge polygons in permafrost. *Geological Society of America*, Special Publication 70.

Lauriol, B., and Godbout, L. 1988: Les terrasses de cryoplanation dans le nord du Yukon: distribution, genèse et âge. *Géographie Physique et Quaternaire* 42, pp. 303-14.

Lewkowicz, A.G. 1987: Nature and importance of thermokarst processes, Sand Hills Moraine, Banks Island, Canada. *Geografiska Annaler* 69A, pp. 321-7.

Lewkowicz, A.G. 1989: Periglacial systems. In: D. Briggs, P. Smithson, and T. Ball, *Physical Geography* (Copp, Clark, Pitman, Toronto), pp. 363-97.

Liebling, R.S., and Scherp, H.S. 1983: Systematic unequal dissection of opposing valley slopes. *Journal of Glaciology* 29, pp. 512-14.

Liverman, D.G.E., Catto, N.R., and Rutter, N.W. 1989: Laurentide glaciation in west-central Alberta: a single (late Wisconsin) event. *Canadian Journal of Earth Sciences* 26, pp. 266-74.

Luckman, B.H. 1976: Rockfalls and rockfall inventory data: some observations from Surprise Valley, Jasper National Park, Canada. *Earth Surface Processes* 1, pp. 287-98.

Luckman, B.H. 1977: The geomorphic activity of snow avalanches. *Geografiska Annaler* 59A, pp. 31-48.

Luckman, B.H. 1981: The geomorphology of the Alberta Rocky Mountains. A review and commentary. *Zeitschrift für Geomorphology* Supplement Band 37, pp. 91-119.

Luckman, B.H., and Crockett, K.J. 1978: Distribution and characteristics of rock glaciers in the southern part of Jasper National Park, Alberta. *Canadian Journal of Earth Sciences* 15, pp. 540-50.

McCann, S.B. ed. 1980: *The Coastline of Canada*, Geological Survey of Canada, Paper 80-10.

McCann, S.B., and Bryant, E.A. 1972: Barrier islands, sand spits and dunes in the southern Gulf of St Lawrence. *Maritime Sediments* 8, pp. 104-6.

McCann, S.B., Dale, J.E., and Hale, P.B. 1981: Subarctic tidal flats in areas of large tidal range, southern Baffin Island, eastern Canada. *Géographie Physique et Quaternaire* 35, pp. 183-204.

Mackay, J.R. 1966: Pingos in Canada. In: Permafrost International Conference, *Proceedings of the Academy of Science – National Research Council*, Publication 1287, pp. 71-6.

Mackay, J.R. 1972: The world of underground ice. *Annals of the American Association of Geographers* 62, pp. 1-22.

Mackay, J.R. 1980: The origin of hummocks, western

Arctic coast, Canada. *Canadian Journal of Earth Sciences* 17, pp. 996-1006.

Mackay, J.R. 1989: Canadian landform examples – 14 Ice-wedge cracks, western Arctic coast. *Canadian Geographer* 33, pp. 365-8.

McKeague, J.A., Grant, D.R., Kodama, H., Beke, G.J., and Wang, C. 1983: Properties and genesis of a soil and the underlying gibbsite-bearing saprolite, Cape Breton Island, Canada. *Canadian Journal of Earth Sciences* 20, pp. 37-48.

MacLennan, M.J. 1988: Canadian landform examples – 8 The Holleford meteorite impact crater. *Canadian Geographer* 32, pp. 173-7.

McRoberts, E.C., and Morgenstern, N.R. 1974: The stability of thawing slopes. *Canadian Geotechnical Journal* 11, pp. 447-69.

Martini, I.P. 1981a: Ice effect on erosion and sedimentation on the Ontario shores of James Bay, Canada. *Zeitschrift fur Geomorphologie* 25, pp. 1-16.

Martini, I.P. 1981b: Coastal dunes of Ontario: distribution and geomorphology. *Géographie Physique et Quaternaire* 35, pp. 219-29.

Mathews, W.H., and McTaggart, K.C. 1969: The Hope Landslide, British Columbia. *Geological Association of Canada, Proceedings* 20, pp. 65-75.

Menzies, J. 1979: A review of the literature on the formation and location of drumlins. *Earth Science Reviews* 14, pp. 315-59.

Menzies, J. 1989: Drumlins – products of controlled or uncontrolled glaciodynamic response? *Quaternary Science Reviews* 8, pp. 151-8.

Mokievsky-Zubok, O. 1977: Glacier-caused slide near Pylon Peak, British Columbia. *Canadian Journal of Earth Sciences* 14, pp. 2657-62.

Mollard, J.D. 1977: Regional landslide types in Canada. In: D.R. Coates (ed.), *Landslides* (Geological Society of America, Boulder, Colorado), Reviews in Engineering Geology III, pp. 29-56.

Monger, J.W.H., and Price, R.A. 1979: Geodynamic evolution of the Canadian Cordillera – progress and problems. *Canadian Journal of Earth Sciences* 16, pp. 770-91.

Moore, D.P., and Mathew, W.H. 1978: The Rubble Creek Landslide, southwestern British Columbia.

Canadian Journal of Earth Sciences 15, pp. 1039-52.

Moran, S.R., Clayton, L., Hooke, R. LeB., Fenton, M.M., and Andriashek, L.D. 1980: Glacier-bed landforms of the prairie region of North America. *Journal of Geology* 25, pp. 457-76.

Morisawa, M. 1968: *Streams* (McGraw-Hill, New York).

Mugridge, S.-J., and Young, H.R. 1983: Disintegration of shale by cyclic wetting and drying and frost action. *Canadian Journal of Earth Sciences* 20, pp. 568-76.

Nanson, G.C., and Hickin, E.J. 1986: A statistical analysis of bank erosion and channel migration in western Canada. *Geological Society of America Bulletin* 97, pp. 497-504.

Neill, C.R. 1965: Measurement of bridge scour and bed changes in a flooding sand-bed river. *Proceedings of the Institute of Civil Engineers (U.K.)* 30, pp. 415-36.

Neill, C.R. 1969: Bed forms in the lower Red Deer River, Alberta. *Journal of Hydrology* 7, pp. 58-65.

NRC 1969: Papers given at a glacial surge seminar and at a symposium (National Research Council). *Canadian Journal of Earth Sciences* 6, pp. 807-1018.

Occhietti, S. 1983: Laurentide ice sheet: oceanic and climatic implications. *Palaeogeography, Palaeoclimatology, Palaeoecology* 44, pp. 1-22.

Odynsky, W. 1958: U-shaped dunes and effective wind directions in Alberta. *Canadian Journal of Soil Science* 38, pp. 56-62.

Ollier, C.D. 1969: *Weathering* (Oliver and Boyd, Edinburgh).

Owens, E.H. 1976: The effects of ice on the littoral zone at Richibucto Head, eastern New Brunswick. *Revue de Géographie de Montréal* 30, pp. 95-104.

Owens, E.H., and Bowen, A.J. 1977: Coastal environments of the Maritime Provinces. *Maritime Sediments* 13, pp. 1-31.

Owens, E.H., and McCann, S.B. 1980: The coastal geomorphology of the Magdalen Islands, Quebec. In: S.B. McCann (ed.), *The Coastline of Canada*, Geological Survey of Canada, Paper 80-10, pp. 51-72.

Parent, M., and Occhietti, S. 1988: Late Wisconsin deglaciation and Champlain Sea invasion in the St Lawrence Valley. *Géographie Physique et Quaternaire* 42, pp. 215-46.

Parizek, J. 1969: Glacial ice-contact rings and ridges. In: S.A. Schumm and W.C. Bradley (eds), *United States Contributions to Quaternary Research*, Geological Society of America, Special Publication 123, pp. 49-102.

Parkes, J.G.M., and Day, J.C. 1975: The hazard of sensitive clay – a case study of the Ottawa-Hull area. *Geographical Review* 65, pp. 198-213.

Pearce, A.J. 1976: Contemporary rates of bedrock weathering, Sudbury, Ontario. *Canadian Journal of Earth Sciences* 13, pp. 188-93.

Peltier, L.C. 1950: The geographic cycle in periglacial regions. *Annals of the Association of American Geographers* 40, pp. 214-36.

Pethick, J. 1984: *An Introduction to Coastal Geomorphology* (Edward Arnold, London).

Ponton, J.R. 1972: Hydraulic geometry in the Green and Birkenhead basins, British Columbia. In: H.O. Slaymaker and H.J. McPherson (eds), *Mountain Geomorphology: Geomorphological Processes in the Canadian Cordillera*, British Columbia, Geographical Series, 14, pp. 151-60.

Prest, V.K. 1968: Nomenclature of moraines and ice-flow features as applied to the Glacial Map of Canada. *Geological Survey of Canada*, Paper 67-57.

Prest, V.K. 1970: Quaternary geology of Canada. In R.J.W. Douglas (ed.), *Geology and Economic Minerals of Canada*, Geological Survey of Canada, Economic Geology Report 1, pp. 675-764.

Prest, V.K. 1984: The late Wisconsin glacier complex. In: R.J. Fulton (ed.), *Quaternary Stratigraphy of Canada – a Canadian Contribution to IGCP Project 24*, Geological Survey of Canada, Paper 84-10, pp. 21-36.

Quigley, R.M., Zajic, J.E., McKyes, E., and Yong, R.N. 1973: Biochemical alteration and heave of black shale; detailed observations and interpretations. *Canadian Journal of Earth Sciences* 10, pp. 1005-15.

Quinlan, G. 1984: Postglacial rebound and the focal mechanisms of eastern Canadian earthquakes. *Canadian Journal of Earth Sciences* 21, pp. 1018-23.

Ringrose, S. 1982: Depositional processes in the development of eskers in Manitoba. In: R. Davidson-Arnott,

W. Nickling, and B.D. Fahey (eds), *Glacial, Glaciofluvial and Glaciolacustrine Systems* (Geo Books, Norwich), 6th Guelph Symposium in Geomorphology, pp. 117-35.

Ritter, D.F. 1978: *Process Geomorphology* (Wm. C. Brown, Dubuque, Iowa).

Rosen, P.S. 1979: Boulder barricades in central Labrador. *Journal of Sedimentary Petrology* 49, pp. 1113-24.

Ryder, J.M. 1971a: Some aspects of the morphometry of paraglacial alluvial fans in south-central British Columbia. *Canadian Journal of Earth Sciences* 8, pp. 1252-64.

Ryder, J.M. 1971b: The stratigraphy and morphology of para-glacial alluvial fans in south-central British Columbia. *Canadian Journal of Earth Sciences* 8, pp. 279-98.

Ryder, J.M. 1981: Geomorphology of the southern part of the Coast Mountains of British Columbia. *Zeitschrift fur Geomorphology* Supplement Band 37, pp. 120-47.

Ryder, J.M., and Church, M. 1986: The Lillooet terraces of Fraser River: a palaeoenvironmental enquiry. *Canadian Journal of Earth Sciences* 23, pp. 869-84.

Schroeder, J., and Arseneault, S. 1978: Discussion d'un karst dans le gypse d'Hillsborough, Nouveau-Brunswick. *Géographie Physique et Quaternaire* 32, pp. 249-61.

Scott, J.S. 1976: Geology of Canadian tills. In: R.F. Legget (ed.), *Glacial Till*, Royal Society of Canada, Special Publication 12, pp. 50-66.

Selby, M.J. 1985: *Earth's Changing Surface* (Oxford University Press, Oxford).

Seppala, M. 1988: Rock pingos in northern Ungava Peninsula, Québec, Canada. *Canadian Journal of Earth Sciences* 25, pp. 629-34.

Shackleton, N.J., and Opdyke, N.D. 1973: Oxygen isotope and palaeomagnetic stratigraphy of equatorial Pacific core V28-238: oxygen isotope temperatures and ice volumes on a 10^5 and 10^6 year scale. *Quaternary Research* 3, pp. 39-55.

Shaw, J. 1988: Subglacial erosion marks, Wilton Creek, Ontario. *Canadian Journal of Earth Sciences* 25, pp. 1256-67.

Shaw, J., and Kvill, D. 1984: A glaciofluvial origin

for drumlins of the Livingstone Lake area, Saskatchewan. *Canadian Journal of Earth Sciences* 21, pp. 1442-59.

Shaw, J., and Sharpe, D.R. 1987: Drumlin formation by subglacial meltwater erosion. *Canadian Journal of Earth Sciences* 24, pp. 2316-22.

Sherstone, D.A. 1983: Sediment removal during an extreme summer storm: Muskwa River, northeastern British Columbia. *Canadian Geotechnical Journal* 20, pp. 329-35.

Shilts, W.W. 1980: Flow patterns in the central North American ice sheet. *Nature* 286, pp. 213-18.

Shilts, W.W., Aylsworth, J.M., Kaszycki, C.A., and Klassen, R.A. 1987: Canadian Shield. In: W.L. Graf (ed.), *Geomorphic Systems of North America*, Geological Society of America, Centennial Special Volume 2, pp. 119-61.

Shilts, W.W., Cunningham, C.M., and Kaszycki, C.A. 1979: Keewatin Ice Sheet – re-evaluation of the traditional concept of the Laurentide ice sheet. *Geology* 7, pp. 537-41.

Slaymaker, O., and McPherson, H.J. 1977: An overview of geomorphic processes in the Canadian Cordillera. *Zeitschrift fur Geomorphologie* 21, pp. 169-86.

Smalley, I.J. 1984: Canada's silty soils – loess in disguise? *Geos* 1984/1, pp. 20-1.

Smith, D.G. 1973: Aggradation of the Alexandra-North Saskatchewan River, Banff Park, Alberta. In: M. Morisawa (ed.), *Fluvial Geomorphology* (George Allen and Unwin, London), pp. 201-19.

Smith, D.G. 1983: Anastomosed fluvial deposits: modern examples from western Canada. In: J.D. Collinson and J. Lewin (eds), *Modern and Ancient Fluvial Systems* (Blackwell, Oxford), International Association of Sedimentologists, Special Publication 6, pp. 155-68.

Smith, D.G., and Smith, N.D. 1980: Sedimentation in anastomosed river systems: examples from alluvial valleys near Banff, Alberta. *Journal of Sedimentary Geology* 50, pp. 157-64.

Smith, D.J. 1987a: Frost heave activity in the Mount Rae area, Canadian Rocky Mountains. *Arctic and Alpine Research* 19, pp. 155-66.

Smith, D.J. 1987b: Solifluction in the southern Canadian Rockies. *Canadian Geographer* 31, pp. 309-18.

Smith, N.D. 1974: Sedimentology and bar formation in the upper Kicking Horse River, a braided outwash stream. *Journal of Geology* 82, pp. 205-23.

Stearn, C.W. 1975: Canada. In R.W. Fairbridge (ed.), *Encyclopedia of World Regional Geology* (Dowden, Hutchinson and Ross, Stroudsburg, Penn.), pp. 139-44.

Stearn, C.W., Carroll, R.L., and Clark, T.H. 1979: *The Geological Evolution of North America* (John Wiley, New York), 3rd ed.

Stene, L.P. 1980: Observations on lateral and overbank deposition – evidence from Holocene terraces, southwestern Alberta. *Geology* 8, pp. 314-17.

Stichling, W. 1974: Sediment loads in Canadian rivers. Water Resources Branch, Environment Canada, *Inland Waters Directorate*, Technical Bulletin 74.

Stockwell, C.H. 1982: Proposals for time classification and correlation of Precambrian rocks and events in Canada and adjacent areas of the Canadian Shield. Part 1: a time classification of Precambrian rocks and events. *Geological Survey of Canada*, Paper 80-19.

Sugden, D.E. 1976: A case against deep erosion of shields by ice sheets. *Geology* 4, pp. 580-2.

Sugden, D.E. 1978: Glacial erosion by the Laurentide ice sheet. *Journal of Glaciology* 20, pp. 367-92.

Sugden, D.E., and John, B.S. 1976: *Glaciers and Landscape* (Edward Arnold, London).

Swift, D.J.P., and Borns, H.W. Jr. 1967: A raised fluviomarine outwash terrace, north shore of the Minas Basin, Nova Scotia. *Journal of Geology* 75, pp. 693-710.

Taylor, R.B., and McCann, S.B. 1983: Coastal depositional landforms in northern Canada. In: J.T. Lowe and A. Dawson (eds), *Shorelines and Isostasy* (Academic Press, London), Institute of British Geographers, Special Publication 16, pp. 53-75.

Taylor, R.B., Wittmann, S.L., Milne, M.J., and Kober, S.M. 1985: Beach morphology and coastal changes at selected sites, mainland Nova Scotia. *Geological Survey of Canada*, Paper 85-12.

Teller, J.T. 1987: Proglacial lakes and the southern margin of the Laurentide ice sheet. In: W.F. Ruddiman and H.E. Wright (eds), *North America and Adjacent Oceans During the Last Deglaciation*, Geological Society of America, The Geology of North America, K-3, pp. 39-69.

Teller, J.T., and Clayton, L. 1983: Glacial Lake Agassiz. *Geological Society of Canada*, Special Paper 26.

Tinkler, K.J. 1986: Canadian landform examples – 2 Niagara Falls. *Canadian Geographer* 30, pp. 367-371.

Trenhaile, A.S. 1975: The morphology of a drumlin field. *Annals of the Association of American Geographers* 65, pp. 297-312.

Trenhaile, A.S. 1976: Cirque morphometry in the Canadian Cordillera. *Annals of the Association of American Geographers* 66, pp. 451-62.

Trenhaile, A.S. 1978: The shore platforms of Gaspé, Québec. *Annals of the Association of American Geographers* 68, pp. 95-114.

Trenhaile, A.S. 1987: *The Geomorphology of Rock Coasts* (Oxford University Press, Oxford, U.K.).

Trettin, H.P., and Balkwill, H.R. 1979: Contributions to the tectonic history of the Innuitian Province, Arctic Canada. *Canadian Journal of Earth Sciences* 16, pp. 748-69.

Tsui, Po C., Cruden, D.M., and Thomas, S. 1989: Ice-thrust terrains and glaciotectonic settings in central Alberta. *Canadian Journal of Earth Sciences* 26, pp. 1308-18.

VanDine, D.F. 1985: Debris flows and debris torrents in the southern Canadian Cordillera. *Canadian Geotechnical Journal* 22, pp. 44-68.

Varnes, D.J. 1978: Slope movement types and processes. In: *Landslides: Analysis and Control* (Transport Research Board, National Academy of Sciences, Washington, DC), Special Report, pp. 11-33.

Vincent, J.-S., and Prest, V.K. 1987: The early Wisconsinan history of the Laurentide ice sheet. *Géographie Physique et Quaternaire* 41, pp. 199-213.

Wang, Y., and Piper, D.J.W. 1982: Dynamic geomorphology of the drumlin coast of southeast Cape Breton Island. *Maritime Sediments and Atlantic Geology* 18, pp. 1-27.

Watts, S.H. 1983: Weathering processes and products under arid Arctic conditions. *Geografiska Annaler* 65A, pp. 85-98.

Watts, S.H. 1985: A scanning electron microscope study of bedrock microfractures in granites under high Arctic conditions. *Earth Surface Processes and Landforms* 10, pp. 161-72.

White, W.A. 1972: Deep erosion by continental ice sheets. *Geological Society of America Bulletin* 83, pp. 1037-56.

White, W.A. 1988: More on deep glacial erosion by continental ice sheets and their tongues of distributary ice. *Quaternary Research* 30, pp. 137-50.

Whitham, K. 1975: The estimation of seismic risk in Canada. *Geoscience Canada* 2, pp. 133-40.

Wigley, T.M.L., Drake, J.J., Quinlan, J.F., and Ford, D.C. 1973: Geomorphology and geochemistry of a gypsum karst near Canal Flats, British Columbia. *Canadian Journal of Earth Sciences* 10, pp. 113-29.

Williams, H. 1979: Appalachian orogen in Canada. *Canadian Journal of Earth Sciences* 16, pp. 792-807.

Windley, B.F. 1984: *The Evolving Continents* (John Wiley, Chichester), 2nd ed.

Woo, M.-K., and Sauriol, J. 1980: Channel development in snow-filled valleys, Resolute, NWT, Canada. *Geografiska Annaler* 62A, pp. 37-56.

Wright, L.D., and Short, A.D. 1984: Morphodynamic variability of surf zones and beaches: a synthesis. *Marine Geology* 56, pp. 93-118.

Zoltai, S.C., and Tarnocai, C. 1981: Some nonsorted patterned ground types in northern Canada. *Arctic and Alpine Research* 13, pp. 139-51.

Index

Page numbers in italic type denote illustrations; some terms not listed here may be found in the Glossary.

THE
GEOMORPHOLOGY
OF CANADA

AN INTRODUCTION

ALAN S. TRENHAILE

Intended as a basic text for junior undergraduates and a continuing source for senior students. *The Geomorphology of Canada* presents a systematic explanation of Canada's landforms, with particular reference to its unique legacy of glacial and periglacial activity. Individual chapters discuss such topics as weathering, mass movement, rivers, coastlines, and karst. Illustrated with over 100 graphs and charts, the book includes an extensive glossary explaining terms that may not be familiar to beginning students.

Alan S. Trenhaile, Professor of Geography at the University of Windsor, is also the author of *The Geomorphology of Rock Coasts* (Oxford, 1987).

OXFORD UNIVERSITY PRESS

ISBN 0-19-540791-1

9 780195 407914